PRAISE FOR EMF DETOX WORKBOOK

#2 in the *Detox Deep Dive* series

"Sophia Ruan Gushée's *EMF Detox Workbook* has done the world a great favor by providing an invaluable resource for anyone concerned about why and how to reduce exposures to electromagnetic fields with understandable explanations of the complex ways that invisible EMFs affect our DNA, and our risk of developing cancer, reproductive and brain disorders, poor sleep quality, emotional well-being, and more.

Detailing how to enjoy technology and stay safer, this guide provides clear, compelling, concise, and practical strategies to avoid or lessen exposures. Buy, use, and share this workbook with your family, friends, colleagues, and schools. They will thank you."

—Dr. DEVRA DAVIS, founder and president of Environmental Health Trust and Nobel Peace Prize winner

"As the leader of a global fitness brand, and a father of two young kids, healthy living is top of mind, always. If you're worried about how technology and toxins are affecting your family, Sophia Ruan Gushée's *Detox Deep Dive* series provides the tools and solutions you need. Her *Home Detox Workbook* shares simple ways to enjoy life with fewer toxic chemicals and heavy metals, and her *EMF Detox Workbook* is must-read as it empowers us with practical strategies to reduce radiation exposure from our technology. I cannot recommend them highly enough!"

—JOEY GONZALEZ, CEO of Barry's Bootcamp

"This well informed offering that Sophia gives us helps to bring awareness to issues we may not collectively be able to feel the effects of yet, but will be a forefront of wellness in the years to come. Because of that, I say she is ahead of her time."

—DEBORAH HANEKAMP, founder of Space by Mama Medicine and author of *Ritual Baths: Be Your Own Healer*

"Actualizing an informed and prudent approach to remediating exposure to the human-made electromagnetic field radiation that occurs in every home has proven to result in improved health and overall wellness. Understanding and taking responsible control of the hidden chemical and electromagnetic toxins in your home will empower you and ensure that you and your family have the best chance to thrive.

Often, simple, cost-free, mindful changes to our daily habits and how we use our electrical devices can produce many noticeable benefits. If you follow the steps of Sophia Ruan Gushée's *EMF Detox Workbook*, with a calm and persistent plan, you may notice results within weeks, if not sooner. If you do not see improvement, you may have more structural issues with your home, which could be causing actionable electromagnetic radiation.

For this, you need a Building Biology® Institute trained and certified EMRS, Professional Electromagnetic Radiation Specialist.™ An EMRS can identify possible stray current entering the utilities in your home, electrical wiring issues, radiative emissions from cell transmitters, inductive emissions from your building's electrical system, or nearby power lines. Once identified and isolated, she or he can provide a myriad of good, best, and better solutions.

Over the past couple of years, I have gotten to know Sophia quite well. I have become genuinely impressed by her steadfast earnestness to learn how to nurture her family by developing a remediation process with balanced and thoughtful tenacity.

I am proud to know someone who has compassion and generous concern for her community by providing this accessible and invaluable *EMF Detox Workbook*. Please read it, practice it, benefit from it, and then pay-it-forward to give back to your extended family and community."

—JAMES FINN, founder of Elexana, LLC, an electromagnetic solutions company serving most major engineering firms operating in the USA

"Two of our three kids have significant medical needs that make them much more sensitive and vulnerable to their environment. We noticed that our youngest often struggled with too much iPad time. Our daughter is non-verbal and relies on technology, often connected to her wrist for rapid communication. We were concerned about the impact of WiFi and Bluetooth technology in relationship with their neurological needs. Their neurologist strongly advocated limiting exposure to Bluetooth technology to avoid causing further complications to their nervous system and seizure disorder.

It was one thing to minimize our children's direct exposure to WiFi and Bluetooth by keeping all technology on airplane mode, but quite another to really understand and then minimize the risks within our home. Sophia's *EMF Detox Workbook* is practical and user-friendly. As a busy mom, it gave me a quick overview and education of risks and hot spots that exist within our home. I found the tips and suggestions easy to incorporate, and they helped reduce a lot of stress and anxiety I had about wanting to minimize our family's exposure to EMFs but not knowing where to start. *EMF Detox Workbook* is an outstanding guide for anyone looking to make tangible changes with minimal disruption and stress."

— COLLEEN OLSON, co-founder and president of DHPS Foundation, serving those with rare genetic disorders

EMF DETOX WORKBOOK

Checklists to Recover from Electromagnetic Exposure

By Sophia Ruan Gushée

Copyright © 2020 Sophia Ruan Gushée
EMF Detox Workbook: Checklists To Recover From Electromagnetic Exposure
All rights reserved.

No part of this publication may be reproduced, distributed, or transmitted in any form or by any means, including photocopying, recording, or other electronic or mechanical methods, without the prior written permission of the publisher, except in the case of brief quotations embodied in critical reviews and certain other non-commercial uses permitted by copyright law. For requests, please email inquiries@nontoxicliving.tips.

Sophia Ruan Gushée

The S File Publishing, LLC

First Printing September 2020
First Edition September 2020
ISBN-13: 978-0-9911401-4-5

Book cover design by Christy Collins, Constellation Book Services.

AFFILIATES

Sophia Ruan Gushée and related entities earn, or may earn, affiliate fees if you purchase things from web links on her websites. As of this printing, these affiliates include Amazon and Intellipure. Please find the most updated list of affiliate partners at https://www.nontoxicliving.tips.

DISCLAIMER

The information in this document, or in any referenced materials, is a general discussion about home, health, and related subjects; and should not be construed as a substitute for professional or medical expertise. The information contained in these topics is provided for educational purposes only, may not be comprehensive, and may become outdated as science and recommendations by experts change.

Please use critical thinking when caring for your home, technology, and other belongings. The author accepts no responsibility or liability for damages that may result from anything in this document or referenced materials. The reader must assess the risks and rewards of pursuing any changes from content herein and referenced.

If you or any other person has a medical concern, consult with your trusted healthcare provider or seek other professional medical treatment. Never disregard professional medical advice, or delay seeking it because of something you have read in this document or in any related content at, or by, the D-Tox Academy, *Practical Nontoxic Living* podcast, https://www.NontoxicLiving.tips, https://www.RuanLiving.com, or Sophia Ruan Gushée.

The opinions and views expressed in this document and related content have no relation to those of any academic, hospital, health practice, or other institutions.

ADDITIONAL WORKS BY SOPHIA RUAN GUSHÉE

Below are additional resources to support your practical nontoxic and healing journey.

BOOKS

- *A to Z of D-Toxing: The Ultimate Guide to Reducing Our Toxic Exposures*
- *Home Detox Workbook: Checklists to Eliminate Toxic Chemicals*
- *EMF Detox Workbook: Checklists to Recover from Electromagnetic Exposure*
- *My Detox Playbook / My Healing Playbook* (for clients and special workshops)

PODCAST

- *Practical Nontoxic Living*

WORKSHOPS & PROGRAMS WITH THE D-TOX ACADEMY

The D-Tox Academy at https://www.nontoxicliving.tips offers various gateways into, and pillars to support, practical nontoxic living and healing. You can start with the most budget-friendly essentials through Home Detox and EMF Detox. Or, dive into other dimensions, like cleaning, self-care, diet, home design, and children's stuff. For more personalized guidance and support, check out the Ruan Detox Immersion Program at https://www.ruanliving.com.

- Detox 101
- Home Detox
- EMF Detox
- Cleaning Detox
- Self-Care Detox

NEWSLETTER

Text "DETOX" to 66866 to tune into practical nontoxic living through Sophia's free email newsletter.

USE YOUR SUPPORT

for members of the online EMF Detox at the D-Tox Academy

This *EMF Detox Workbook* was designed for the online EMF Detox workshop at the D-Tox Academy. The online EMF Detox offers "power lessons" (videos that provide context for these checklists, which can make it easier to detox with others that you live with); a members-only online forum to ask questions and learn from others; and relevant video interviews with guests of the *Practical Nontoxic Living* podcast.

Please remember that the *EMF Detox Workbook'* checklists are numbered in connection to how the online videos are numbered so that you can use both the online and offline resources more easily. However, this workbook can be very effective even independent of the online EMF Detox workshop.

To log in:

- ❑ Visit https://www.NontoxicLiving.tips. Once there, click on "Login" in the upper-right corner of the screen.
- ❑ Interact with the D-Tox Academy more conveniently via an app. Just download the "Kajabi" app onto your phone or tablet to get started.

To start:

- ❑ Check into your member portal, Pillar 1.
- ❑ Watch the EMF Detox Introduction video in Pillar 3.

To onboard others:

- ❑ Watch the online D-Tox Academy videos with those you live with (no matter how young they are). This encourages cooperation and starts needed conversations for healthy technology habits. These videos and ensuing conversations promote self-awareness for optimal energy hygiene.

NOT YET A MEMBER?

As an owner of this workbook, enjoy a **special offer** when purchasing the online EMF Detox workshop.

- ❑ Learn more at https://www.NontoxicLiving.tips/emf-detox-workbook.

TABLE OF CONTENTS

INTRODUCTION 9

1. EMF D-TOX APPROACH 17

2. HEALTHY TECHNOLOGY HABITS 33

3. HOME EMF EDIT 45

4. EVENING EMF EDIT 63

5. BODY EMF EDIT 69

6. CONGRATULATIONS! 79

7. MY SOCIAL INSPIRATIONS 81

8. NOTES 83

For your easy reference, fill in the rest of this table of contents with your key topics in Chapter 8, and their respective starting page numbers. Use pencil or erasable pen to maximize your flexibility to edit.

Observations from your elimination diet 85

....

....

....

....

....

....

APPENDIX

Select References 106

Detox Deep Dive 108

INTRODUCTION

Create recovery times from your EMF exposure.

Before June 29, 2007, most people had never heard of an iPhone. However, upon its invention and release, it quickly and dramatically transformed our lives.

Now, over 13 years later, we have 7+ *billion* cell phones in the world (Gushée 2020), and countless other wireless devices. In America, an average household is estimated to have 11 "connected" devices (Spangler 2019).

And this is expected to continue to grow at a phenomenal pace, partially due to the Internet of Things (IoT), a technology advancement that will enable many more products, including *household* products, to communicate wirelessly. The IoT is projected to contribute to 50 billion more devices by 2030 (Gushée 2020), offering us "smart" homes, cars, wearables, and more.

As our use of wireless devices has advanced—from speaking through telephones to listening to music on mp3 files, to watching movies from any internet-enabled device, to video conferencing with people from around the world—so has the technology infrastructure that allows us to access even more data in less time.

More specifically: 2G, 3G, 4G, and 5G are generations of network infrastructures that have allowed—or will allow—more data to be exchanged far more efficiently than ever before. For example, a 2-hour video would take 26 hours to download on 3G, 6 minutes on 4G, and 3.5 seconds on 5G (Hoffman 2020).

We live in a sea of man-made radiation.

While this brings invaluable opportunities, it also requires a lot of different *types* of energy—aka, radiation, or man-made electromagnetic fields (EMFs)—to transmit vast amounts of data within seconds.

As a result, we live in a sea of radiation in our workplaces, public spaces, schools, and homes. Our bodies and brains endure *unprecedented* radiation exposure all day and all night.

Peer-reviewed studies show harmful effects at our current EMF exposure levels.

Some of the world's leading scientists—including over 236 reputable scientists from 41 countries who have peer-reviewed publications on EMFs—and 1,000+ health professionals have been urging stronger laws to protect human health. They determined that the scientific evidence proves serious risks from common wireless radiation exposure, e.g., peer-reviewed studies associate radiation from our household wireless devices (like cell phones) with cancer and damage to DNA, reproductive organs, and the brain. (Bioinitiative Report 2012, Environmental Health Trust 2020d, Hippocrates Electrosmog Appeal Belgium, Kalmbacher 2020)

The most extensive review of the science on this topic is the Bio-Initiative Report, which shares the analysis of 1,800 peer-reviewed scientific studies on possible health risks from wireless technologies and electromagnetic fields. It results from the collaboration of 29 independent scientists and health experts from around the world.

Authors include three former presidents of the Bioelectromagnetics Society, the Chair of the Russian National Committee on Non-Ionizing Radiation, a Senior Advisor to the European Environmental Agency, ten with medical degrees (MDs), and 21 PhDs. With supporting details on its website (https://bioinitiative.org/), it concludes:

> Bioeffects are clearly established and occur at very low levels of exposure to electromagnetic fields and radiofrequency radiation. Bioeffects can occur in the first few minutes at levels associated with cell and cordless phone use. Bioeffects can also occur from just minutes of exposure to mobile phone masts (cell towers), WI-FI, and wireless utility 'smart' meters that produce whole-body exposure.

In the proceeding paragraphs, I will refer to the peer-reviewed scientific EMF health studies as the "Data."

EMF exposure will not necessarily cause symptoms or illness.

People are often afraid to learn about toxic exposures. It can be empowering to remember that toxic exposures do not necessarily cause illness. They just increase your risks. So we want to manage risks. For example, cancer-causing exposures do not necessarily cause cancer, but they increase your chances of developing cancer. Lifestyle and environmental factors can help avoid, reduce, or repair harm from toxic exposures.

How our modern EMF exposure affects us is complex as there are many factors at play. Key considerations that make us vulnerable to toxic chemicals, which I explain in my book *A to Z of D-Toxing: The Ultimate Guide to Reducing Our Toxic Exposures*, also influence our vulnerability to electromagnetic fields. For example, our health and well-being result from the interplay of various factors, such as our genes, environments, lifestyle choices, and stress. While further study is underway, we know that EMF considerations include the types of radiation we are exposed to, our duration and frequency of EMF exposure, and when in life we were, or are, exposed.

Effects can manifest immediately, over a lifetime, and over generations. They also may not be noticeable to us.

> Many of these bioeffects can reasonably be presumed to result in adverse health effects if the exposures are prolonged or chronic. This is because they interfere with normal body processes (disrupt homeostasis), prevent the body from healing damaged DNA, produce immune system imbalances, metabolic disruption and lower resilience to disease across multiple pathways. Essential body processes can eventually be disabled by incessant external stresses (from system-wide electrophysiological interference) and lead to pervasive impairment of metabolic and reproductive functions. (Bioinitiative Report 2012)

Decades of research are needed before the effects of our EMF exposure can be understood more fully. However, with an informed approach, which the *EMF Detox Workbook* provides, we can enjoy less harm and more recovery.

Governments' responses to the Data have been proactive, reactive, and inactive.

When scientific evidence shows that the status quo is risky but there is still scientific uncertainty about alternatives, how should governments respond?

The "precautionary" approach, central in European Union environmental law, is a decision-making strategy that manages complex risks with the perspective that it is better to be safe than sorry. (Science for Environment Policy 2017) I view this as proactive.

Activated by the Data, countries that have enacted protective measures from EMFs include Belgium, Cyprus, European Parliament, France, Israel, Spain, and Russia (Environmental Health Trust 2020a, Environmental Health Trust 2020c). The Parliamentary Assembly of the Council of Europe, the parliamentary arm of the Council of Europe, warned in 2011:

> Given the context of growing [EMF] exposure of the population, in particular that of vulnerable groups such as young people and children, there could be extremely high human and economic costs if early warnings are neglected... Waiting for high levels of scientific and clinical proof before taking action to prevent well-known risks can lead to very high health and economic costs, as was the case with asbestos, leaded petrol and tobacco.

Hence, countries—like Cyprus, France, Israel, and Russia—have restricted WiFi from schools and pursued other public health measures. Recently, the Russia Ministry of Health banned smartphones for education and recommends books instead of computers for digital learning. (Environmental Health Trust 2020c)

In the US, reaction to the Data is gradually occuring. Some states have taken protective measures, such as Connecticut, Maryland, Oregon and California (Environmental Health Trust 2020c). In 2014, the California Medical Association helped by passing a Wireless Resolution that warned:

> peer reviewed research has demonstrated adverse biological effects of wireless EMF including single and double stranded DNA breaks, creation of reactive oxygen species, immune dysfunction, cognitive processing effects, stress protein synthesis in the brain, altered brain development, sleep and memory disturbances, ADHD, abnormal behavior, sperm dysfunction, and brain tumors.

However, federal action has not been motivated by the Data. For example, the US Food and Drug Administration (2020) website says:

> Based on the evaluation of the currently available information, the FDA believes that the weight of scientific evidence has not linked exposure to radio frequency energy from cell phone use with any health problems at or below the radio frequency exposure limits set by the FCC.

At a national level, the US remains inactivated by the Data. As a result, the public endures unnecessary risk. Too many historical examples demonstrate that even after proof of harm is established, regulatory change in America comes slowly.

US EMF safety standards are based on +24-year-old tests.

In America, the Federal Communications Commission (FCC) determines the emissions' limits of radio frequency radiation by cell phones and similar wireless products. Current wireless emission standards were established in 1996 and have not been updated since despite urgent requests from scientific and medical communities.

> Even though no scientific evidence currently establishes a definite link between wireless device use and cancer or other illnesses... some consumers are skeptical of the science and/or the analysis that underlies the FCC's RF exposure guidelines (US Federal Communications Commission 2019, accessed July 2020).

It was hard to reconcile the FCC position with others. However, in the critically-acclaimed book *Disconnect: The*

Truth About Cell Phone Radiation, What the Industry Is Doing to Hide It, and How to Protect Your Family, author and National Book Award finalist Dr. Devra Davis provides a convincing explanation.

From writing my chapter titled "Track Records of Manipulative Business Strategies" in my book *A to Z of D-Toxing: The Ultimate Guide to Reducing Our Toxic Exposures*, I recognized strategies Dr. Davis describes by the telecommunications industry as ones used by the tobacco industry. They were so effective that other industries, like asbestos, implemented them too to delay regulation. In the broader section titled "Track Records" in *A to Z of D-Toxing*, I recount how litigation was instrumental in protecting public health from tobacco and asbestos.

Because of this background, I was relieved by two lawsuits filed in early 2020.

TWO LAWSUITS MAY SAVE US.

Two of the world's preeminent experts on how EMFs threaten human health are challenging the FCC to update its EMF standards through two separate lawsuits.

ENVIRONMENTAL HEALTH TRUST.

One of these public health heroes is Dr. Devra Davis, an epidemiologist who has earned much praise and many awards including the 2007 Nobel Peace Prize, which she won alongside Vice President Al Gore.

Dr. Davis has pioneering expertise on how public health is affected by environmental exposures, including wireless radiation. She founded the scientific, nonprofit think-tank, the Environmental Health Trust (EHT) in 2007 to promote "a healthier environment through research, education, and policy" and to create a "thriving world where technology is both state-of-the-art and safe for all," according to its website.

In early 2020, the EHT filed a lawsuit against the FCC. With support from the Natural Resources Defense Council, one of the nation's top environmental organizations, and joined by other scientists and citizens, EHT contends that the FCC has failed to protect public health by ignoring hundreds of peer-reviewed studies and the advice of the American Academy of Pediatrics, and others; and, instead, continues to rely on "24-year-old safety tests designed when phones were the size of a shoe and used by few" (Kalmbacher 2020).

Dr. Davis is uniquely qualified to beget stronger EMF regulations—not only through her epidemiological expertise but also because of her impressive career in- and out- of government. Learn more about Dr. Davis and how EMFs can affect us by listening to episode #24 of the *Practical Nontoxic Living* podcast.

CHILDREN'S HEALTH DEFENSE.

The second world-renowned scientist contesting FCC guidelines is Dr. David O. Carpenter. Dr. Carpenter is part of a lawsuit filed by the Children's Health Defense, a non-profit organization, led by its Chairman Robert F. Kennedy, Jr. Dedicated to ending the epidemic of children's chronic health conditions, the Children's Health Defense "recognizes that wireless technology radiation is a major contributing factor to the exponential increase in sickness among children" (Children's Health Defense 2020).

A graduate of Harvard Medical School, Dr. Carpenter is Director of the Institute for Health and Environment at the University of Albany, a collaborating center for the World Health Organization. Another public health hero,

Dr. Carpenter has a distinguished record as a public health physician and advocate. Among his many impressive accomplishments is Dr. Carpenter's role as co-editor of the Bio-Initiative Report (introduced earlier). Read an excerpt from one of my *Practical Nontoxic Living* podcasts with Dr. Carpenter after this Introduction (Box 1).

WE HAVE UNIQUE VULNERABILITIES.

As a mother of three young kids, I am deeply grateful for these lawsuits as they create a more foreseeable opportunity for safer EMF standards. Each lawsuit claims that the FCC guidelines ignore extensive peer-reviewed studies that prove harmful health effects at levels of wireless radiation that are lower than the current FCC guidelines. Inevitably, these lawsuits will also raise public awareness and society's protection of the most vulnerable.

WE UNDERGO VULNERABLE STAGES OF LIFE.

Young life, especially prenatally, is particularly vulnerable to EMFs. One key reason is that our biological foundation starts forming in utero, and continues for decades under the complex influence of many things, including our genes, hormones, environmental exposures, and lifestyle choices. In children, their fast-growing—and immature—biological system is one way they are more vulnerable to environmental exposures, including EMFs.

For example, the brain is most vulnerable during its first 20+ years. Dr. Devra Davis explains in episode #24 of my *Practical Nontoxic Living* podcast that the brain grows to 100 billion cells from conception until birth, and then doubles within the first 2 years of life. The brain continues to develop into our 20s, and a protective barrier—the blood-brain barrier (BBB)—that protects the adult brain from toxic compounds in the body is not yet fully developed in children. Furthermore, children's brains absorb more radiation than adults due to their smaller heads, thinner skulls and "higher conductivity of the brain tissue" (European Environment Agency 2019).

Dr. Davis explains in this same episode that wireless radiation can weaken the BBB (including in adults as well as the membrane of sperm), rendering the brain (and sperm) more vulnerable to a number of risks. But those who use cell phones before age 20 have 4 times more brain cancer by the time they reach adulthood.

ELECTROMAGNETIC SENSITIVITY VARIES.

Just as we experience unique sensitivities to common exposures—like dairy and nuts—some people suffer from electromagnetic sensitivity. As we are complex beings that rely on biochemical and bioelectrical processes, humans experience a spectrum of tolerances. Some people: appear unaffected; have chronic symptoms; develop a terminal or chronic illness; and some are disabled by EMF exposures. Terms used to describe symptoms that worsen with EMF exposure (and often improve with reduced EMF exposure) are electromagnetic sensitivity or "electromagnetic hypersensitivity" (EHS).

Historically, EHS has been rarely accepted as a medical diagnosis. Increasingly, however, as scientific and medical experts urge more understanding and acknowledgment of EHS (and multiple chemical sensitivity (MCS)), EHS has become recognized by organizations, governments, and even legal systems around the world (Dewey 2015).

> In view of our present scientific knowledge, we thereby stress all national and international bodies and institutions, more particularly the World Health Organization (WHO), to recognize EHS and MCS as true medical conditions which acting as *sentinel diseases* may create a major public health concern in

years to come worldwide...Inaction is a cost to society and is not an option any more.

—Brussels International Scientific Declaration on EHS and MCS, 2015

Examples of organizations that accept EHS include the Americans with Disabilities Act, the US Access Board (United States Access Board 2006), and the State of New York Public Service Commision (State Of New York Public Service Commission 2019). In the United States, the Governors of Alabama, Colorado, and Connecticut have issued proclamations that recognize electromagnetic sensitivity. (Environmental Health Trust 2020b)

Legal systems around the world are ruling on behalf of those with EHS. For example, in 2015, a *Washington Post* article, titled "Are 'WiFi allergies' a real thing? A quick guide to electromagnetic hypersensitivity" by Caitlin Dewey, reported that the French legal system determined that a 39-year-old woman was eligible for disability benefits from her EHS (nearly $900/month). The article shed light on the growing international acknolwedgement of EHS:

> The French ruling also adds to a growing body of global legal precedent on EHS and its legitimacy. Courts in Australia have already awarded workers' comp to EHS patients. In Sweden, the syndrome is officially classified as a "functional impairment," which affords sufferers a range of legal protections and accommodations.

Peer-reviewed scientific evidence has found (Brussels International Scientific Declaration 2015):

- a high and growing number of people worldwide are suffering from EHS and MCS;
- EHS and MCS affect women, men and children;
- many frequencies of the electromagnetic spectrum (including extremely low frequencies) and multiple chemicals are involved in the occurrence of EHS and MCS respectively;
- the trigger for illness can be from acute high intensity exposure or chronic very low intensity exposure;
- reversibility can be obtained by reducing exposure to man-made EMFs and chemicals;
- EHS and MCS are causing serious consequences to some people's health, professional and family life;
- EHS and MCS should be recognized by institutions with responsibility for human health.

The online EMF Detox and the *EMF Detox Workbook* help you today as lawsuits take years to unfold. Furthermore, laws do not empower you with the framework provided by the EMF D-Tox Approach, which informs you of the accessible choices you have to detox your EMFs.

EMF ELIMINATION DIET.

Decreasing EMF exposure reduces risks for serious health issues that take a long time to manifest. But people experience benefits in the shorter term too. For example, many sleep and feel better relatively soon. To discover your unique reactions, I invite you to experiment with an EMF elimination diet.

Incorporating reported EHS symptoms, Chapter 1 lists health issues that EMF exposures can, or may, contribute to. The *EMF Detox Workbook* leads you through an elimination of EMFs, and asks you to observe your reactions.

> The primary method of treatment [for EMF-related health issues] should mainly focus on the prevention or reduction of EMF exposure...at home and at the workplace.
>
> —European Academy for Clinical Environmental Medicine EMF Guideline 2016 for the Prevention, Diagnosis and Treatment of EMF-related Health Problems and Illnesses.

The EMF D-Tox Approach (Chapter 1) informs you of the EMF exposure you can control, and asks you to use a "Love Test" to decide which EMF exposure you will not mind eliminating. What survives the Love Test are your worthwhile "Risks."

CREATE RECOVERY TIMES.

This workbook outlines practical ways both to eliminate unnecessary EMF exposure and to establish healthy boundaries—recovery times—from your Risks. This can help your resiliency for the EMF exposure you cannot control. For example, by reducing EMF exposure when sleeping, you can increase your sleep quality and promote restorative rest for your brain and body. This promotes resiliency for the EMF exposure outside your influence.

As you work through the checklists, the EMF D-Tox Approach will empower you with strategies to harness chronic, imposing, man-made energy to create periods of reduced EMFs, i.e., recovery times. Especially at home.

DEVELOP HEALTHY ENERGY HYGIENE.

By learning how to create frequent "vacations" from your digital and energetic accessibility, the EMF D-Tox Approach leads you towards sustainable habits that minimize unnecessary risks from EMF exposure while also protecting your joy. This "energy hygiene" not only decreases physical risks but also enhances the conditions for improved sleep quality, fertility and pregnancy outcomes, energy, attention, cognitive skills, and presence. When you are digitally disconnected, you can nourish your reconnections with yourself and others—upgrading your overall physical, social, emotional, and mental well-being. Being mindful and protective of your attention, and what you see/read/hear can trigger a cascade of countless benefits.

Regardless of how quickly behavioral changes come, awareness is invaluable in optimizing your trajectory. Establishing healthy technology habits will be essential for overall well-being. Thank you for including this workbook in your journey!

INSPIRED BY
TBS

Sincerely,

Sophia Ruan Gushée

* *Simplifying a highly complex topic, this introduction was based on years of research, Practical Nontoxic Living podcast interviews, and personal experience. The most helpful references and works cited are in Select References in the Appendix.*

BOX 1.

Are you or your loved ones as resilient as a 154-pound, white male?

Below is a snippet from my *Practical Nontoxic Living* podcast with Dr. David O. Carpenter, which provides insight into the complexities of our EMF exposures, why FCC guidelines need to be updated to consider unique vulnerabilities, and why it is wise to avoid unnecessary EMF exposures.

Sophia: Would you talk about the testing of safe radiation levels? I've read that it's historically been conducted on a model that is based on an adult man about 6-feet tall and 200 pounds, therefore not considering the unique vulnerabilities of children, fetuses and maybe women.

Dr. Carpenter: In the past, everything was focused on a 70-kilogram white male. 70-kilogram white males are not even the majority of the population!

> *many people—whether they're a fetus or an old person or a 70-kilogram man—may have a genetic susceptibility so that they're more vulnerable than the average person*

Then finally, people realized, well, women aren't necessarily responding the same as men. They have different body structures; they have different hormone structures. Then more recently, we realized that the fetus and young children are very much more vulnerable to a lot of things than adults are. And also, that the elderly are more vulnerable because their immune systems have been declining; they have multiple health issues. Then finally I think we've begun to address those issues.

But the other issue that has not really been addressed is genetic susceptibility. So many people—whether they're a fetus or an old person or a 70-kilogram man—may have a genetic susceptibility so that they're more vulnerable to some exposure than the average person. So we've gone from trying to protect the average adult white male to then understanding women and young children and the elderly need separate protection.

But the question is, when the government sets a standard, should it be for the average person or should it be for the most vulnerable person? The way it's been in the past, usually there's a safety factor or fudge factor. So if you have clear evidence that a certain concentration of a chemical causes cancer, you would go down in the order of magnitude or two orders of magnitude in concentration for your standard. That is to presumably protect the more vulnerable person.

But I think increasingly we're going to find, as we understand more about human genetics, that what is responsible is to try to protect the most vulnerable members of society, not a 70-kilo white male.

—Dr. David O. Carpenter, MD, Director of the Institute for Health and the Environment, a Collaborating Center of the World Health Organization, in *Practical Nontoxic Living* podcast, episode #15

—1—

EMF D-TOX APPROACH

The EMF D-Tox Approach protects your joy and well-being.

Grounded by five cornerstones, the EMF D-Tox Approach raises awareness to your chronic EMF exposures as it helps you identify the EMF exposures you will not miss and can easily reduce. As you eliminate EMF exposures, you can record improvements you notice. This helps develop your radar for sensing when you should detox your EMF exposures.

Cornerstones.

The checklists in this workbook lead you to apply the five cornerstones of the EMF D-Tox Approach (below) to your home and lifestyle as is relevant.

1. Observe your EMF "elimination diet."
2. Conduct the Love Test.
3. Establish "Energy Recovery Opportunities."
4. Evolve your mindset for a lifetime of optimized energy hygiene.
5. Celebrate your progress!

Intention.

After applying this EMF D-Tox Approach with the guidance of the checklists (even better: enhance your experience with the online EMF Detox workshop), you will have gained enough experience to use your intuitive common sense to avoid EMF exposure.

By pursuing this EMF D-Tox Approach as an elimination diet of EMFs, you may notice your body's responses to certain EMF exposures (like Cellular, WiFi, or Bluetooth). Eventually, you may be able to "listen" to your body as it reacts to EMFs. As it communicates with you through symptoms, sensations, and sleep quality, you will develop a better understanding of how you can modify your EMF exposure to take care of yourself.

OBSERVE YOUR ELIMINATION DIET

CORNERSTONE 1.

Some experts say we are all sensitive to EMFs since human beings (and other life forms) are bioelectrical systems. However, most people are not aware of how they are influenced by EMFs.

Given how chronic our EMF exposures are, everyone would benefit from experimenting from an EMF elimination diet, during which a participant observes improvements, if any, after removing one EMF exposure at a time. This is the best way to learn how you react to EMF exposures. Similar to the idea that some people are sensitive to nuts (or another exposure), some are also sensitive to EMFs, which is explained in the Introduction.

Cornerstone 1 involves two steps.

- *Step 1: Create your Baseline Assessment.* The next page lists symptoms or health issues reported or proven to be associated with electromagnetic exposures. Review them and circle the issues that are relevant to you for a baseline assessment or to recognize how this EMF D-Tox Approach can reduce the risks of your circled health concerns. I will refer to your results as your **"Baseline Assessment."**
    - Please note that the symptoms on the next page can result from various other factors too. And the list is not intended to be comprehensive or to diagnose. But it can raise your awareness and curiosity to your responses to EMF exposures.
    - Visit the following websites to review additional health effects that EMF exposures may contribute to:
        - Bio-Initiative Report at https://bioinitiative.org/conclusions/.
        - Environmental Health Trust at https://ehtrust.org/.
        - Baby Safe Project at https://www.babysafeproject.org/.
    - From reviewing the websites above, consider recording (in Chapter 8 Notes, or elsewhere in this workbook) motivating data that can help you or your loved ones continue trying to change habits. The Select References in the Appendix also offer excellent resources from which to learn more.
- *Step 2: Observe and record.*
    - Use Chapter 8 Notes to list the health issues that are relevant to you.
    - Use Chapter 8 Notes to **track your observations** as you pursue this EMF D-Tox Approach.

Document your Baseline Assessment.

❑ **Circle** the issues below that are relevant to you for your Baseline Assessment.

❑ **List** your health issues in Chapter 8 Notes.

- acoustic neuroma
- ADD
- ADHD
- acid reflux
- adrenal failure
- autism
- birth defects
- brain damage
- brain fog
- brain and nervous system tumors
- brain: compromised blood-brain barrier
- brain: an uncomfortable feeling of heat or pressure inside the head
- breast cancer
- breast cancer medications rendered ineffective
- cataracts (premature)
- cellular health: chromosome breaks, suppression of DNA mechanisms, DNA mutation, suppressed calcium ion channeling, cellular inflammation, mitochondrial damage
- chronic fatigue
- concentration difficulties
- confusion
- depression
- dermatological symptoms (redness, tingling, burning sensations, unexplained rashes)
- diabetes
- digestive disturbances
- dizziness
- dry eyes, sinuses, or throat
- fibromyalgia
- GERD
- heart arrhythmia, heart palpitation, increased heart rate
- high blood pressure
- high triglyceride levels
- high glucose levels
- immune system imbalances
- infertility
- insomnia
- intestinal lining weakened
- irritability
- leukemia
- leaky gut syndrome
- learning disorders
- lymphoma
- melanoma
- memory loss (short-term)
- migraine headaches
- miscarriage
- mood swings
- muscle pain
- muscle weakness
- nausea
- neurological disorders
- noise sensitivity
- obesity
- placenta (damaged or weakened)
- slow reflexes
- sperm: low sperm count
- sperm: low sperm motility
- tinnitus
- tiredness
- unexplained infertility
- vertigo
- others

LOVE TEST

CORNERSTONE 2.

The *EMF Detox Workbook* will teach you about your opportunities to reduce your EMF exposure from what you own, buy, and do. Guided by the checklists, you will follow the three steps of the "**Love Test.**"

> ## "LOVE TEST"
>
> As you discover your products and habits that pose health risks, ask yourself, *Do I love it? Do I need it?*
>
> Given the health risks you have learned, ask yourself, *Is it worth it anyway?*
>
> If no, then avoid (eliminate, phase out, or don't buy) this EMF exposure.
>
> If yes, then keep it for now and move onto other detox tweaks that feel easy to pursue.
>
> I call this the "Love Test." This EMF D-Tox Approach protects your joy and helps you detox with ease.*
>
> *Please note that the Love Test and EMF D-Tox Approach were designed for the average healthy person. If you have serious health issues, consult your healthcare providers on whether a more aggressive detox of toxic exposures should be pursued.*

Worthwhile risks.

In the rest of this workbook, I will refer to the things and habits that survive the Love Test as "**Risk.**" These Risks have been deemed worthwhile or beloved by you!

With your Risks, wonder how you can reduce your EMF exposures from them. For example, if you love your fitness watch that emits EMF exposures 24/7, would you not mind disabling EMF exposures during your sleep? Or, would you not mind wearing the fitness watch just during your workouts? Cornerstone 3 helps you decrease your EMF exposure from your Risks—what you love, need, or find worthwhile.

In summary.

The Love Test will guide you to:

1. Audit your shopping, things, and habits for EMFs (this is to raise your awareness to the EMF exposures you can influence).
2. Conduct the Love Test. What passes the Love Test are your worthwhile Risks.
3. Edit your shopping, things, and habits accordingly.

Cornerstone 3 helps you reduce your risks from what passes the Love Test.

ENERGY RECOVERY OPPORTUNITIES

CORNERSTONE 3.

In Cornerstone 3 of our EMF D-Tox Approach, we will explore ways to detox your beloved Risks through the six strategies that follow. I will refer to them collectively as "Energy Recovery Opportunities."

ENERGY RECOVERY OPPORTUNITIES

Enjoy your Risks with less EMF exposure. To achieve this, the EMF Detox checklists will guide you to apply the six strategies below when relevant. Exercising applying these strategies will provide excellent experience from which you can use your common sense to decrease EMF exposures even beyond those listed in the *EMF Detox Workbook*. As you implement these Energy Recovery Opportunities more often, your body gains more chances to thrive as it was brilliantly designed to do.

1. **"Minimize Time."** Minimize your time near the EMF source. By being intentional with your EMF exposure, you can benefit from the EMF source when you need it or when the Risks are worthwhile to you. But by choosing your EMF exposure mindfully, you are empowered to avoid EMF exposure when the Risks are not worthwhile to you.
2. **"Maximize Distance."** Maximize the distance between you and the EMF source because EMFs dissipate with distance.
3. **Create EMF Naps.** Create EMF Naps by disabling EMFs for nap time and enabling them for wake times. I will refer to this disabled time as an "**EMF Nap**." To create EMF Naps for your technology:
    a. *Disable* Cellular/WiFi/Bluetooth connectivity (I will refer to this connectivity as "**Wireless Emissions**") and/or electrical connectivity (I will call this "**Electrical Emissions**") for the EMF Nap, and enable them for the wake time. To disable and enable:
        i. Manually:
            1. Power off devices; or,
            2. Use airplane mode to disble Cellular/WiFi/Bluetooth. "**Airplane Mode**" will mean the disabling of Cellular/WiFi/Bluetooth connectivity.
        ii. Or, automatically, which is discussed in the checklists.
    b. Disable Electrical Emissions by:
        i. Unplugging electrical products; or,
        ii. Using products that Ground the Electricity (strategy 5 below).

continued on the next page

4. **"Disable and Cord."** This refers to the opportunity to disable Wireless Emissions and still enjoy the internet-dependent technology with a wired connection via an Ethernet cable.

5. **"Ground the Electricity**." This refers to using certain products (like a timer, 3-prong power strip, or grounding adapters) to reduce, or stop, the flow of electricity demanded by a product plugged into an electrical outlet.

6. **"Detox the Technology Settings**." With some product(s), you can disable Wireless Emissions in the products' settings. Track your notes in Chapter 8.

You will see these six Energy Recovery Opportunities (in bold font above as well as the Defined Terms on the next page) recommended throughout the checklists when relevant.

DEFINED TERMS

You will notice the terms in this section throughout the checklists. They are meant to reinforce key concepts that can help you develop an intuitive understanding of how you may detox the countless EMF sources from what you buy, own, and do.

Airplane Mode. Using airplane mode on your wireless devices to disable Cellular, WiFi, and/or Bluetooth connectivity and emissions.

Audit. Observing or taking inventory of your habits or products for the purpose of avoiding EMF exposure that you will not miss to mindfully limit your EMF exposure. This is step 1 of the Love Test.

Baseline Assessment. The health issues that you want to track or reduce the risks of. This results from Cornerstone 1.

Detox the Technology Settings. Reducing a product's EMF emissions by disabling its Wireless Emissions in the product's settings. This is introduced in Cornerstone 3 of the EMF D-Tox Approach (Chapter 1), and strategy six of the six strategies to enjoy Energy Recovery Opportunities.

Disable and Cord. Reducing a product's EMF emissions by disabling its Wireless Emissions and still enjoying the internet-dependent technology with an Ethernet cable. This is introduced in Cornerstone 3 of the EMF D-Tox Approach (Chapter 1), and strategy four of the six strategies to enjoy Energy Recovery Opportunities.

Electrical Emissions. Unhealthy or risky energy created from products connected to electricity. Most often, these products are plugged into an electrical outlet or power strip. They can include "dirty electricity" too. This is introduced in Cornerstone 3 of the EMF D-Tox Approach (Chapter 1), and strategy three of the six strategies to enjoy Energy Recovery Opportunities.

EMF. An acronym for electromagnetic fields. In the *EMF Detox Workbook*, EMF is an umbrella term for diverse types of energy created from wireless and wired technologies. These types of energy include (but are not limited to) ionizing radiation, non-ionizing radiation, and dirty electricity. In this workbook, EMF includes the defined terms: Wireless Emissions and Electrical Emissions.

EMF D-Tox Approach. An EMF-reduction strategy that includes five cornerstones introduced in Chapter 1: (1) observe your EMF elimination diet; (2) conduct the Love Test; (3) establish "Energy Recovery Opportunities"; (4) evolve your mindset; and (5) celebrate your progress! The main goal of the EMF D-Tox Approach is to empower you to reduce EMFs you will not miss so you can increase your opportunities to recover from the EMFs that are unavoidable.

continued on the next page

EMF Nap. Creating mindful EMF exposure by disabling its Wireless Emissions and/or Electrical Emissions when you are not enjoying benefit(s) from them, and enabling those emissions when you find the Risks worthwhile. This is introduced in Cornerstone 3 of the EMF D-Tox Approach (Chapter 1), and strategy three of the six strategies to enjoy Energy Recovery Opportunities.

Energy Recovery Opportunities. Six strategies of the EMF D-Tox Approach that can reduce EMF exposure from your worthwhile Risks. Introduced in Chapter 1 (Cornerstone 3), these six strategies include: 1) Minimize Time; 2) Maximize Distance; 3) Create EMF Naps; 4) Disable and Cord; 5) Ground the Electricity; and 6) Detox the Technology Settings.

Ground the Electricity. Reducing a product's Electrical Emissions. This may be accomplished by plugging 2- or 3- pronged products into a timer, 3-prong power strip, or grounding adapter. These products can reduce, or stop, Electrical Emissions. This is introduced in Cornerstone 3 of the EMF D-Tox Approach (Chapter 1), and strategy five of the six strategies to enjoy Energy Recovery Opportunities.

Love Test. A test introduced in Cornerstone 2 of the EMF D-Tox Approach (Chapter 1) that raises awareness to one's EMF exposures and related risks from one's products and habits. After becoming aware of these EMF exposures and risks, one decides if the EMF exposure is worthwhile.

Maximize Distance. Reducing one's EMF exposure by maximizing distance from the EMF source since EMFs dissipate with distance. This is introduced in Cornerstone 3 of the EMF D-Tox Approach (Chapter 1), and strategy two of the six strategies to enjoy Energy Recovery Opportunities.

Minimize Time. Reducing one's EMF exposure by minimizing one's time near an EMF source. This is introduced in Cornerstone 3 of the EMF D-Tox Approach (Chapter 1), and strategy one of the six strategies to enjoy Energy Recovery Opportunities.

Risk. Products, habits, and other choices that survive the Love Test.

Tech Home. A technology charging station introduced in checklist 3.9 Tech Home.

Tech House Rules. Guidelines for the Tech Home that create not only EMF recovery times but also social, emotional, and psychological benefits as well. Discussed in checklist 3.9 Tech Home.

Wireless Emissions. Radiation from Cellular, WiFi, and/or Bluetooth connectivity. This is introduced in Cornerstone 3 of the EMF D-Tox Approach (Chapter 1), and strategy three of the six strategies to enjoy Energy Recovery Opportunities.

EVOLVE YOUR MINDSET

CORNERSTONE 4.

Change is always hard: learning new things, modifying behavior, and adjusting your choices while living with others who may or may not be trying to evolve too. Inevitably, your process of growth will include highs and lows.

The key to long-term improvement is the right mindset. The seven guidelines below can influence your mindset to focus on your many successes to come, regardless of when things are not "perfect."

MINDSET FOR A LIFETIME OF OPTIMIZED ENERGY HYGIENE

7 Guidelines to Enjoy a Sustainable EMF Detox

1. Detox at your unique pace.
2. Hold onto what you love and need.
3. Start with higher impact areas, "low hanging fruit" (what is easy), or what feels non-threatening.
4. Do not judge yourself or others. Just observe.
5. Take time—and practice patience and forgiveness—to incorporate one detox tweak at a time into a new habit.
6. Celebrate your progress.
7. Repeat all of the above as is comfortable for you.

LISTEN & LEARN

Recommended Practical Nontoxic Living podcasts

- ❏ #24: "Protect Your Brain and Body from 5G and Other EMFs" with Dr. Devra Davis, Nobel-winning scientist and author of *Disconnect: The Truth About Cell Phone Radiation*.

- ❏ #22: "Heal Your Eyes From Digital Screens" with Dr. Marc Grossman, author of *Natural Eye Care: Your Guide to Healthy Vision and Healing*.

- ❏ #17: "EMF Protection Tips from a Former Telecommunications Engineer" with Daniel DeBaun, author of *Radiation Nation: The Fallout of Modern Technology*.

- ❏ #15: "Headaches, Nausea, and Fatigue. Might You Be Electrohypersensitive?" with Dr. David O. Carpenter, MD, co-editor of the BioInitiative Reports (check them out at https://bioinitiative.org).

- ❏ #10: "How Toxic Exposures Threaten Our Reproductive Health and Children's Health" with Dr. Hugh Taylor, Yale School of Medicine. If you look forward to starting or growing your family or have young children. Dr. Taylor talks about his research on the risks of prenatal cell phone radiation exposure.

- ❏ Subscribe to the the *Practical Nontoxic Living* podcast via your favorite podcast platform, such as Apple Podcasts, Google Podcasts, Overcast, Spotify, and Stitcher.

CELEBRATE YOUR PROGRESS!

CORNERSTONE 5.

Inherent in the EMF D-Tox Approach is celebrating all positive changes—no matter how small. This section is meant to help shape your perspective so you are proud of yourself throughout your inevitable non-linear path towards healthy change, i.e., your path would graph like a wave with high and low inflection points.

Feel proud of any efforts to learn because awareness is hard for many people to embrace. But awareness serves as fertilizer for seeds of transformation. Below are suggestions for **recording your highs** in the Chapter 8 Notes section to remind you of your accomplishments for if, or when, you feel moments of disappointment or failure.

Congratulations! By reading this page, you have already:

- ❑ Bought the *EMF Detox Workbook*.
- ❑ Read _____ pages of this book.
- ❑ Learned why detoxing your EMFs is important.

And you might have (check off everything you have done and do not worry about the accuracy of the math, just estimate because the objective is for you to appreciate your progress):

- ❑ Completed your Baseline Assessment.
- ❑ Completed one or more application(s) of the Love Test.
- ❑ Watched _____ EMF Detox power lesson(s) at the D-Tox Academy at https://www.NontoxicLiving.tips.
- ❑ Reduced your cell phone's Wireless Emissions by approximately _____ hours per day by using airplane mode more often. This is an approximate _____% reduction from your weekly exposures.
- ❑ Reduced your WiFi router's Wireless Emissions by approximately _____ hours per night.
- ❑ This is an approximate _____% reduction from your weekly exposures.
- ❑ Reduced your fitness accessories' Wireless Emissions by approximately _____ hours per week by using it only when you find it worthwhile.
- ❑ Listened to some *Practical Nontoxic Living* podcasts that are recommended in these checklists.
- ❑ Made incredible progress! Just skim this workbook to see all the healthy tips you have learned.

❑ Mark the **checklists** that you have reviewed so far.

❑ 2.1 Airplane Mode	❑ 3.1 Power Strips	❑ 3.9 Tech Home
❑ 2.2 Signal Strength	❑ 3.2 Cordless Phones	❑ 4.1 Blue Light
❑ 2.3 Distance	❑ 3.3 WiFi Routers	❑ 4.2 Your Bedroom
❑ 2.4 Cell Phone Calls	❑ 3.4 Home Entertainment	❑ 5.1 Unique Vulnerabilities
❑ 2.5 Texting	❑ 3.5 Appliances	❑ 5.2 Medical Imaging
❑ 2.6 Laptops	❑ 3.6 Home Office	❑ 5.3 Earthing
❑ 2.7 Accessories	❑ 3.7 Light Bulbs	❑ 5.4 Natural Remedies
❑ 2.8 Vehicles	❑ 3.8 EMF Expert	❑ 5.5 Digital Detoxes

❑ Mark the **videos** in the online EMF Detox workshop that you have watched so far.

❑ What are EMFs?	❑ 2.5 Texting	❑ 3.7 Light Bulbs
❑ Electromagnetic Spectrum	❑ 2.6 Laptops	❑ 3.8 EMF Expert
❑ Are EMFs harmful?	❑ 2.7 Accessories	❑ 3.9 Tech Home
❑ Benefits of Reducing Your EMF Exposure	❑ 2.8 Vehicles	❑ 4.1 Blue Light
❑ 2.1 Airplane Mode	❑ 3.1 Power Strips	❑ 4.2 Your Bedroom
❑ 2.2 Signal Strength	❑ 3.2 Cordless Phones	❑ 5.1 Unique Vulnerabilities
❑ 2.3 Distance	❑ 3.3 WiFi routers	❑ 5.2 Medical Imaging
❑ 2.4 Cell Phone Calls	❑ 3.4 Home Entertainment	❑ 5.3 Earthing
	❑ 3.5 Appliances	❑ 5.4 Natural Remedies
	❑ 3.6 Home Office	❑ 5.5 Digital Detoxes

❑ Write down in this workbook additional videos you watch to support your EMF Detox, and record motivating data.

❑ BONUS: Spread public awareness by completing Chapter 7 My Social Inspirations!

HONOR YOUR SUCCESSES!

No positive change is too small to celebrate!

Elaborate to Cornerstone 5 in the rest of this chapter.

- ❑ **Add your accomplishments** in the space that follows.
- ❑ Feel free to **add sticky notes, and/or staple additional sheets** to record your milestones!
- ❑ Revisit this section, Cornerstone 5, to **remind yourself of your achievements**.

EMF DETOX

EMF D-TOX APPROACH

"Love Test"

D-TOX YOUR ENERGETIC STRESSORS

1. Avoid (eliminate, phase out, or don't buy) what fails the Love Test.
2. Create Energy Recovery Opportunities from what survives the Love Test. This reduces your risks from what you need and love.

In summary.

In essence, these five Cornerstones of the EMF D-Tox Approach will engage you to:

1. Observe your EMF elimination diet.
2. Conduct the Love Test.
3. Create "Energy Recovery Opportunities."
4. Evolve your mindset for a lifetime of optimized energy hygiene.
5. Celebrate your progress!

Practicing the EMF D-Tox Approach with the checklists in the *EMF Detox Workbook* will establish a great foundation of critical thinking and the mindset needed to optimize your energy hygiene.

—2—

HEALTHY TECHNOLOGY HABITS

Be selective about your energy exposures.

Energy takes infinite forms. Most obvious, it manifests as sound (like music), images, and movement. Less obvious are the invisible forms of energy, like EMFs.

The *EMF Detox Workbook*, however, raises your awareness of this invisible energy. And the online videos at the D-Tox Academy are even more helpful. With a simple informational framework, you can start relying on your intuition and common sense to identify Energy Recovery Opportunities.

Inspired by your desire to reduce your EMF exposure, as you establish healthy technology habits, you will notice it creates positive ripple effects. For example, as you use Airplane Mode at dinner, you become more present for those you are having dinner with. As you establish parameters with how you work with your laptop, you will take more breaks, which are good for your eyes and can promote more movement during your day.

Manage your expectations by remembering that perfection is impossible. You will continue to have Airplane Mode off during some dinners, and you will sometimes work too long on your laptop—and with your laptop's Wireless Emissions and Electrical Emissions on. However, your goal is to reduce the number and duration of those times over your lifetime. That is a sustainable detox approach.

2.1 AIRPLANE MODE

When wireless connectivity is not needed on digital devices, it is best to create an EMF Nap by powering OFF our Wireless Emissions. And then powering them ON when Wireless Emissions are useful. However, this behavior is not practical for most people. An alternative to creating an EMF Nap from powering devices on/off is to use Airplane Mode to disable/enable Wireless Emissions more conveniently. In this workbook, "**Airplane Mode**" will assume that Airplane Mode disables Cellular/WiFi/Bluetooth emissions.

LOVE TEST

- ❏ **Audit.** Notice the times your wireless devices create Wireless Emissions.

- ❏ **Conduct the Love Test.** When you notice that your wireless devices (like your cell phone) are not on Airplane Mode (like during the times listed below), ask yourself if the EMF exposure is worthwhile. From your Audit, notice the times you will not mind disabling Wireless Emissions.

- ❏ **Edit your EMF exposure.** With your enhanced awareness to disable Wireless Emissions, choose times to use Airplane Mode. The list below can help. Select the times that feel easy. Revisit the list when you feel ready for more change.

- ❏ **Manage your worthwhile Risks.** Review the Energy Recovery Opportunities below.

CREATE ENERGY RECOVERY OPPORTUNITIES

Use Airplane Mode on your cell phone or other wireless devices:

- ❏ During breakfast.
- ❏ During lunch.
- ❏ During dinner.
- ❏ While exercising.
- ❏ When running errands.
- ❏ While in a vehicle.
- ❏ When you want to be more present, like during a date, when you're with your kids, or while meditating.
- ❏ Before going to bed (keeping it on airplane mode while asleep).
- ❏ Before and during a nap.
- ❏ When children are using, or are around, internet-enabled devices.

- ☐ Around those who are pregnant.
- ☐ While working on a computer.
- ☐ Any other times you do not need the devices' Wireless Emissions.

Confirm:
- ☐ Always confirm that Airplane Mode disabled Cellular/WiFi/Bluetooth. Software updates can modify how devices behave, including Wireless Emissions during airplane mode.

OBSERVE YOUR EMF ELIMINATION DIET
- ☐ As you use Airplane Mode more, record (in Chapter 8 Notes) cause-effect(s) you observe. Many people notice improved sleep quality, and enhanced presence and attention.

EVOLVE YOUR MINDSET
- ☐ Revisit Cornerstone 4 in Chapter 1 to shape your perspective for a lifetime of optimized energy hygiene.

CELEBRATE YOUR PROGRESS!
- ☐ Record your progress in Chapter 1 (Cornerstone 5) to appreciate your growing awareness and achievements. Congratulations, you are off to a great start!

EXCLUSIVE RESOURCES FOR HEALTHY CONVERSATIONS

Each checklist is supplemented by at least one video in the online EMF Detox at the D-Tox Academy. The videos provide context and extra motivation for the checklists. They can be wonderful to watch as a household. Clients have had great conversations with their children by watching a video before dinner and then talking about the video during dinner. Together, they talked about what changes they could pursue.

2.2 SIGNAL STRENGTH

With weak signal strength, as your cell phone searches for a signal, it can emit more EMFs than when it has full signal.

LOVE TEST

- ❏ **Audit.** Check your wireless devices' signal strength when:
    - ❏ Their Wireless Emissions are enabled, like smart speakers that provide virtual assistance.
    - ❏ You are about to use your wireless devices, like a cell phone.
- ❏ **Conduct the Love Test.** Given the greater EMF exposure in areas of weak signal strength, is it worth using your wireless devices when signal strength is low?
- ❏ **Edit your EMF exposure.** After completing the above, revise your sources of Wireless Emissions accordingly.
    - ❏ Skip, or delay, using a wireless device when signal strength is low.
    - ❏ Use a wireless device in areas with stronger signal strength.
    - ❏ Avoid calls in elevators, basements, buses, cars, trains, and the subway. During these times, turn the phone OFF or to Airplane Mode, when possible.
    - ❏ Take precautions extra seriously when near children or pregnant women.
- ❏ **Manage your worthwhile Risks.** Review the Energy Recovery Opportunities below.

CREATE ENERGY RECOVERY OPPORTUNITIES

To reduce EMF exposure when signal strength is low:

- ❏ **Minimize Time.** Minimize the time of your EMF exposure.
- ❏ **Maximize Distance.** Maximize the distance between you and the wireless device when its Wireless Emissions are enabled. For example, if you are using your cell phone for a call, use speakerphone to create distance from your head.
- ❏ **Create EMF Naps.** Disable Wireless Emissions when you have poor service.

2.3 DISTANCE

Radiation exposure decreases significantly with distance from an EMF source.

LOVE TEST

- ❏ **Audit.** Notice the proximity between you and your technology when their Wireless Emissions and/or Electrical Emissions are enabled.
- ❏ **Conduct the Love Test.** Given that EMF exposure increases with proximity to the EMF source, is the proximity worthwhile? Can distance be increased?
- ❏ **Edit your EMF exposure.** After completing the above, revise your distance from EMF sources accordingly.
- ❏ **Manage your worthwhile Risks.** Review the Energy Recovery Opportunities below.

CREATE ENERGY RECOVERY OPPORTUNITIES

To **Maximize Distance** from an EMF source:

- ❏ Avoid keeping a cell phone in your pocket or in a purse very close to your body when Wireless Emissions are on.
- ❏ If you must keep your cell phone near your body (like in a pocket or a purse), disable Wireless Emissions by powering the device off or using Airplane Mode.
- ❏ When on a call, do not place the phone directly against your head. Instead, use a speakerphone or earphones.
    - ❏ Review checklist 2.7 for tips on earphones.
- ❏ When you exercise, do not keep your cell phone close to your body, like in a sports bra or pocket.
    - ❏ If you must, keep your cell phone close to your body during its EMF Nap. For example, you can enjoy music, podcasts, movies, etc. by downloading them in advance so you can enjoy them while Wireless Emissions are disabled.
- ❏ When streaming, downloading, or sending large files, keep the phone away from your head and body.
- ❏ Avoid Bluetooth technology around your head and body (including wireless headphones, Bluetooth stereo headsets, etc.). **Conduct the Love Test** to decrease Bluetooth exposures.

2.4 CELL PHONE CALLS

It is best to minimize the number and duration of your cell phone calls. When you must make calls on cell phones, the tips below will help reduce your exposure to cell phone radiation.

LOVE TEST

- ☐ **Audit.** Observe your habit with cell phone calls.
    - ☐ Notice how much time you spend on cell phone calls.
    - ☐ Notice how much distance there is between you and your cell phone during your calls.
    - ☐ How often do you use earphones (wired or wireless) and speakerphone?
    - ☐ Be mindful of metal since cell phones, and other wireless devices, work harder to get a signal through metal, creating more EMFs.
- ☐ **Conduct the Love Test.** Given the cell phone radiation risk, which cell phone calls are worthwhile? Can the objective of the call be met through the Energy Recovery Opportunities below?
- ☐ **Edit your EMF exposure.** After completing the above, revise the number and duration of your cell phone calls accordingly.
    - ☐ Minimize cell phone calls from cars, elevators, trains, or buses since the metal that create them can exacerbate your EMF exposure.
- ☐ **Manage your worthwhile Risks.** Review the Energy Recovery Opportunities below.

CREATE ENERGY RECOVERY OPPORTUNITIES

When cell phone calls pass the Love Test:

- ☐ **Minimize Time.**
    - ☐ Try to keep your cell phone calls brief.
    - ☐ Schedule more, especially long, conversations for a corded landline. Reference checklist 3.2 Cordless Phones with tips on what to seek in corded phones.
    - ☐ Text instead of call when possible. Reference Checklist 2.5 Texting.
- ☐ **Maximize Distance.**
    - ☐ When you are on a call, maximize the distance between your cell phone and your head, body, and other people.
    - ☐ Use a speakerphone when possible.
    - ☐ Use a headset designed to reduce radiation. Reference Checklist 2.7 Accessories.

OBSERVE YOUR EMF ELIMINATION DIET

- ☐ As you reduce your cell phone calls, record (in Chapter 8 Notes) cause-effect(s) you observe. Many people notice more serenity, and enhanced presence and attention.

EVOLVE YOUR MINDSET

- ☐ Revisit Cornerstone 4 in Chapter 1 to shape your perspective for a lifetime of optimized energy hygiene.

CELEBRATE YOUR PROGRESS!

- ☐ Record your progress in Chapter 1 (Cornerstone 5) to appreciate your growing awareness and achievements. Congratulations, keep returning to this workbook when you are ready for more support or challenge!

2.5 TEXTING

Texting can expose us to less cell phone radiation than making phone calls.

LOVE TEST

- ☐ **Audit.** Notice when you use a wireless device for calls.
- ☐ **Conduct the Love Test.** Given that texting emits less radiation than cell phone calls, when can you text instead of call?
- ☐ **Edit your EMF exposure.** Reduce your cell phone radiation by decreasing your time on cell phone calls and texting more often instead.
- ☐ **Manage your worthwhile Risks.** Review the Energy Recovery Opportunities below.

CREATE ENERGY RECOVERY OPPORTUNITIES

For the times that cell phone calls do not pass the Love Test:

- ☐ Text instead of call, but:
  - ☐ Do not text while driving since this increases the risk of car crashes.
  - ☐ Do not text while walking since "distracted walking" injuries are also on the rise.

2.6 LAPTOPS

We should not rest laptops on our bodies, including our laps. Research has found risks from the heat and radiation emitted from laptops—especially for male sperm, pregnant women, children, and couples trying to conceive.

LOVE TEST

- ❑ **Audit.** Notice when you use your laptop while it is on your body.
- ❑ **Conduct the Love Test.** Given the health risks from working with your laptop on your body, when can you create distance between you and the laptop?
- ❑ **Edit your EMF exposure.** After completing the above, revise how you work with your laptop.
- ❑ **Manage your worthwhile Risks.** Review the Energy Recovery Opportunities below.

CREATE ENERGY RECOVERY OPPORTUNITIES

To decrease your EMF exposure when using a laptop:

- ❑ **Minimize Time.** When you find the Risk of using your laptop on your body worthwhile, minimize this time.
    - ❑ Try to use a wired connection for internet-connectivity. This decreases the times you use your laptops with its Wireless Emissions enabled. Laptops can emit Wireless Emissions from wireless and Bluetooth connectivity.
    - ❑ Be aware that a laptop plugged into an electrical socket emits more Electrical Emissions than when it is not connected to electricity. When you must use your laptop as it is plugged into an electrical outlet, minimize this time.
    - ❑ Reduce your EMF exposures by using the laptop when it is not plugged into an electrical outlet. Therefore, you will have to charge your laptop before you plan on using it. (This is good motivation to take a break from sitting and move your body and eyes.)
- ❑ **Maximize Distance** between you and the laptop.
    - ❑ Avoid putting the laptop on your lap and body (some people put it on their belly when resting on their back).
    - ❑ Charge laptops in a location that is as far from people as possible. Ideally, charge laptops 6-8 feet away from people.

- **Create EMF Naps.**
  - Disable Wireless Emissions and Electrical Emissions as often as possible.
  - Charge your laptop when you are not using it so that you can use it without it being connected to an electrical outlet.
- **Disable and Cord.**
  - Use an Ethernet cable to connect the laptop to the internet whenever possible, and disable WiFi/Bluetooth.
  - Use a wired keyboard and mouse, and disable Bluetooth. Wireless options require Bluetooth technology.
- **Ground the Electricity.** When charging the laptop, use a 3-prong grounding adapter or power strip.
- **Detox the Technology Settings.**
  - Bluetooth is often enabled automatically. Check the settings to disable Bluetooth when it is not needed.
  - Disable WiFi when WiFi is not needed.
  - Decrease Blue light exposures to minimize your sleep disruption. This can be done manually or automatically through your laptop's settings or certain software, like f.lux.

OBSERVE YOUR EMF ELIMINATION DIET

- As you adjust your laptop use, record (in Chapter 8 Notes) cause-effect(s) you observe. Many people notice improved sleep quality, and enhanced presence and attention.

EVOLVE YOUR MINDSET

- Revisit Cornerstone 4 in Chapter 1 to shape your perspective for a lifetime of optimized energy hygiene.

CELEBRATE YOUR PROGRESS!

- Record your progress in Chapter 1 (Cornerstone 5) to appreciate your growing awareness and achievements. Congratulations, you have already learned many high impact tips!

2.7 ACCESSORIES

Various accessories—like earphones, headsets, smart watches, fitness accessories, chargers—and a growing list of innovative products are increasingly incorporating Cellular/WiFi/Bluetooth technology. The safety of Bluetooth has not been thoroughly studied. Some people are sensitive to it. It can be helpful to experiment to see if you feel better with less Bluetooth exposures.

LOVE TEST

- **Audit.** Notice the accessories you have, or want to buy, that create Wireless Emissions. Below are accessories that often create Wireless Emissions.
    - Earphones or headset.
    - Smart watches.
    - Chargers.
    - Fitness accessories.
- **Conduct the Love Test.** Given the health risks from your accessories' Wireless Emissions, which are worth keeping or using?
- **Edit your EMF exposure.** After completing the above, revise which accessories you own.
    - Consider donating or discarding the accessories that fail the Love Test.
- **Manage your worthwhile Risks.** Review the Energy Recovery Opportunities below.

CREATE ENERGY RECOVERY OPPORTUNITIES

For the accessories that pass the Love Test (especially ones that are worn on the body for extended periods of time), explore whether you can modify your relationship with them through the Energy Recovery Opportunities below.

- **Minimize Time.**
    - Earphones and headsets are not radiation-free, and how much radiation they emit may vary. Headsets release small amounts of radio frequency (RF) energy even when phones are not in use, according to the California Department of Public Health. Remove earpieces from ears when they are not in use.
    - Research airtube headsets, which are often recommended as the safest earpieces when considering EMFs. While many people find them less comfortable than other earpieces, better products will become available.
    - Minimize prolonged conversations on a cell phone (even with earphones), and use a landline or text instead.

- **Maximize Distance** between the cell phone and your body, even when using earphones.
- **Create EMF Naps** for your beloved accessories. You can reduce your EMF exposure by reducing the number of hours that accessories (like watches with Wireless Emissions) are worn on the body. When they are unnecessary, Maximize Distance and put them in an EMF Nap.
- **Ground the Electricity** when chargers are plugged into an electrical outlet. Review Cornerstone 3 in Chapter 1 for tips.
- **Detox the Technology Settings.** As you learn how to Detox the Technology Settings of your accessories, work towards changing your habits so your default settings include leaving Wireless Emissions disabled until you need them enabled.

OBSERVE YOUR EMF ELIMINATION DIET
- As you edit your accessories' use, record (in Chapter 8 Notes) cause-effect(s) you observe. Many people notice improved sleep quality, and enhanced presence and attention.

EVOLVE YOUR MINDSET
- Revisit Cornerstone 4 in Chapter 1 to shape your perspective for a lifetime of optimized energy hygiene.

CELEBRATE YOUR PROGRESS!
- Record your progress in Chapter 1 (Cornerstone 5) to appreciate your growing awareness and achievements. Congratulations for editing your accessories!

2.8 VEHICLES

When in a car (or public transportation), the radiation from wireless devices can bounce off of the vehicle's metal shell, intensifying EMF exposures for those within the vehicle.

LOVE TEST
- **Audit.** Notice the sources of Wireless Emissions in your car.
- **Conduct the Love Test.** Given the health risks to these Wireless Emissions, especially in a moving vehicle, is it worthwhile to have their Wireless Emissions enabled?
- **Edit your EMF exposure.** After completing the above, revise which Wireless Emissions are enabled.
- **Manage your worthwhile Risks.** Review the Energy Recovery Opportunities below.

CREATE ENERGY RECOVERY OPPORTUNITIES

To decrease Wireless Emissions in vehicles that pass the Love Test:

- ❏ **Minimize Time.** Enable Wireless Emissions only when you are deriving benefit from them.
    - ❏ When you are not, disable them.
    - ❏ Remember that Wireless Emissions can be stronger when trying to connect in a fast-moving car, bus, or train because it can be harder not to drop calls as they switch connections from one cell tower to the next.
    - ❏ Download music, podcasts, movies, etc. <u>before</u> getting into the vehicle to enjoy them in the vehicle with Wireless Emissions disabled.
    - ❏ Be extra mindful to disable Cellular/WiFi/Bluetooth if there are children or pregnant women in the vehicle.
    - ❏ Check the vehicle to see if the vehicle creates Wireless Emissions (like WiFi and Bluetooth). If so, is it possible to enjoy this vehicle safely with Wireless Emissions disabled?
- ❏ **Maximize Distance.** When you are enjoying your Wireless Emissions, then maximize distance from this EMF source.
- ❏ **Create EMF Naps.** Disable Wireless Emissions of smart devices by either powering OFF the device or through using Airplane Mode (reference checklist 2.1). Wirelessly connect only when necessary or when the Risk is worthwhile.

OBSERVE YOUR EMF ELIMINATION DIET

- ❏ As you modify your EMF sources in vehicles, record (in Chapter 8 Notes) cause-effect(s) you observe. Many people notice improved sleep quality, and enhanced presence and attention.

EVOLVE YOUR MINDSET

- ❏ Revisit Cornerstone 4 in Chapter 1 to shape your perspective for a lifetime of optimized energy hygiene.

CELEBRATE YOUR PROGRESS!

- ❏ Record your progress in Chapter 1 (Cornerstone 5) to appreciate your growing awareness and achievements. Congratulations for knowing how to detox your transportation experience!

— 3 —

HOME EMF EDIT

Filtering and organizing your home's EMFs can help clear your mind and spirit.

With more technology at home than ever, we also see more disorganized cables. As you learn about ways to reduce your EMF exposures, you will naturally clear this clutter and feel more tranquility.

3.1 POWER STRIPS

Many people do not realize that plugged-in appliances produce EMFs (Electrical Emissions can extend 6-8 feet)—even if they are turned off!

LOVE TEST

- ☐ **Audit.** Notice the chargers/cables in your home.
- ☐ **Conduct the Love Test.** Given the health risks from Electrical Emissions, which ones are worth unplugging, or consolidating and grounding with power strips and grounding adapters?
- ☐ **Edit your EMF exposure.** After completing the above, revise your charging/cable situation.
- ☐ **Manage your worthwhile Risks.** Review the Energy Recovery Opportunities below.

CREATE ENERGY RECOVERY OPPORTUNITIES

To reduce your power strips' Electrical Emissions:

- ☐ **Minimize Time.** You can not always be 6-8 feet from your activated power strips, so minimize the time that you are near them.
- ☐ **Maximize Distance.**
  - ☐ Whether turned on or off, plugged in power strips can create Electrical Emissions. Since Electrical Emissions can extend 6 to 8 feet, try to create at least 6 feet (or as much as you can) from plugged-in electrical cords and power strips, including your feet.
  - ☐ Remember that the Electrical Emissions can extend through walls, floors, and ceilings.
- ☐ **Create EMF Naps** for the power strip:
  - ☐ Turn off the power strip at night.
  - ☐ Or connect the power strip to a timer for automated EMF Naps. (This helps save on utility bills as well!).
  - ☐ Or leave the power strip in an EMF Nap, wake only when needed.
- ☐ **Ground the Electricity** from your power strips.
  - ☐ Buy adapters that can reduce, or shut off, Electrical Emissions. You can see which ones I use at my curated Amazon storefront, which is found at https://NontoxicLiving.tips.
  - ☐ Choose 3-prongs when possible since 2-prongs tend to generate more Electrical Emissions.

SOPHIA'S SELECTION

The grounded outlet adapter I use to reduce, or stop, Electrical Emissions can be seen online in two places at https://NontoxicLiving.tips. First, in lesson 3.1 in the online EMF Detox workshop at the D-Tox Academy, I share a brief video on why this is my selection. Second, linked on my website is my curated Amazon store titled, "Nontoxic Living on Amazon," from where you can find the adapter I use.

3.2 CORDLESS PHONES

Cordless phones can be a significant and chronic EMF source, depending on the model. Generally, corded telephones create less Wireless Emissions and Electrical Emissions than cordless phones. They also can provide more reliable and secure communication service, and are useful when cell phone connectivity is interrupted. The tips below are precautionary measures to reduce EMFs.

LOVE TEST

- **Audit.** Notice the cordless phones in your home.
- **Conduct the Love Test.** Given the health risks from their Wireless Emissions and Electrical Emissions, which ones are worth keeping?
- **Edit your EMF exposure.** After completing the above, revise which cordless phones you keep.
    - Consider removing all cordless phones and replacing them with corded landlines that offer speakerphone capabilities and are battery-powered.
    - If you are not ready to eliminate them completely, consider:
        - Researching lower-emitting EMF cordless phones. Improved models should enter the market over time.
        - Having at least one corded phone in your home to help support the transition to less time on cordless phones and to be accessible at night when you unplug your cordless phones (and disable your cell phone).
- **Manage your worthwhile Risks.** Review the Energy Recovery Opportunities below.

CREATE ENERGY RECOVERY OPPORTUNITIES

To reduce your EMF exposure from cordless phones that survive the Love Test:

- **Minimize Time.**
    - Keep conversations short on cordless phones, and use the speakerphone function to Maximize Distance between the phone and your head.
    - Take extra precautions to limit children's exposure to cordless phones, including teenagers.
- **Maximize Distance.**
    - Maximize Distance between the cordless phone base and where people spend lots of time (like the kitchen, family room, and especially where they sleep).
    - Remember that EMFs can travel through walls, floors, and ceilings.
    - Return the handset to the base unit (the charger) when it is not in use. It is emitting radiation when it

is not on that base unit until the battery has run down.

- ❑ **Create EMF Naps** for your cordless phones.
    - ❑ Allow Wireless Emissions and Electrical Emissions only when they serve a purpose (like when you are awake), and disable them at night by unplugging the base unit and removing the batteries from its handset.
    - ❑ To automate the EMF Naps of the cordless phone's base station, consider connecting its power outlet to a timer.
- ❑ **Ground the Electricity.** Can you "ground" your cell phone charging base with a timer, 3- prong power strip, or grounding adapter? This can reduce, or stop, the flow of electricity demanded by the charging station when plugged into an electrical outlet.

OBSERVE YOUR EMF ELIMINATION DIET

- ❑ As you detox your exposure to cordless phones, record (in Chapter 8 Notes) cause-effect(s) you observe. Many people notice improved sleep quality, and enhanced presence and attention.

EVOLVE YOUR MINDSET

- ❑ Revisit Cornerstone 4 in Chapter 1 to shape your perspective for a lifetime of optimized energy hygiene.

CELEBRATE YOUR PROGRESS!

- ❑ Record your progress in Chapter 1 (Cornerstone 5) to appreciate your growing awareness and achievements. Congratulations for detoxing your home's energy!

3.3 WIFI ROUTERS

WiFi routers can create strong, chronic Wireless Emissions in our homes.

LOVE TEST

- ❑ **Audit.** Notice where your WiFi routers are located, and their proximity to where people spend a lot of time.
- ❑ **Conduct the Love Test.** Given the health risks from their Wireless Emissions and Electrical Emissions, are there any that you can disable? Is there a better location for their placement? Are there times you will not mind disabling their Wireless Emissions?
- ❑ **Edit your EMF exposure.** After completing the above, revise which WiFi routers you keep, where they are located, and their EMF Naps.
- ❑ **Manage your worthwhile Risks.** Review the Energy Recovery Opportunities in the next section.

CREATE ENERGY RECOVERY OPPORTUNITIES

To reduce EMF exposure from your WiFi routers:

- **Minimize Time.**
    - Review below how to Create EMF Naps. Your opportunities for recovery increase with more frequent and longer EMF Naps.
- **Maximize Distance** between the WiFi router(s) and where people spend a lot of time, especially where they sleep.
    - Remember that Wireless Emissions can travel through floors and walls. So, for example, an office on the floor below could be an EMF source to a bedroom on the floor above.
- **Create EMF Naps** for your WiFi router. Set up manual or automated ways to disable and enable Wireless Emissions and (maybe) Electrical Emissions at specified times.
    - Make this a part of your evening and morning routines by manually disabling/enabling Wireless Emissions and Electrical Emissions; or,
    - Automate this with a timer to have the WiFi router turn on/off automatically at set times.
    - With some complicated home electronic systems, technicians can program settings to disable/enable the WiFI router automatically. My home has several WiFi routers connected to a "kill switch." The WiFi routers are automatically disabled at bedtime and enabled in the morning. In addition, I can disable/enable them manually during the day with a "kill switch," which was programmed by a technician. *This may affect the settings of devices that depend on WiFi, like music systems.
    - Assess the risks/rewards of your options to identify your preferred strategy.
- **Disable and Cord.** Whenever you can reduce the number of products connecting with your WiFi router(s), then this reduces the EMFs at home. To do this:
    - Use an ethernet cable whenever you can to enjoy your technology (and disable Wireless Emissions).
    - Buy more ethernet cords so you can more easily Disable and Cord.
- **Ground the Electricity.** Can you "ground" your router with a timer, 3-prong power strip, or grounding adapters? This may reduce, or stop (even better from an EMF consideration), the flow of electricity demanded by the router when plugged into an electrical outlet but not needed during an EMF Nap.
- **Detox the Technology Settings.** Depending on the router and your home setup, you may be able to use your WiFi router with Wireless Emissions disabled. Instead, there would be more wired connections created. Professional help is usually needed.
    - Explore the settings of your WiFi router(s) to see if you can reduce the signal strength to a lower level while also establishing sufficient internet/data access and speed.
    - If you do not need WiFi because you have created more wired connections, then disable WiFi on the

WiFi router through its settings. Professional help is often needed.

- ❏ Record (in Chapter 8 Notes) key details for future reference.

Reduce demand of your WiFi router.

- ❏ When using a computer or laptop, take advantage of opportunities to Disable and Cord, Maximize Distance, and Create EMF Naps.
- ❏ For those wireless devices you do not use often, disable Wireless Emissions. Enable them when you need them.
- ❏ Remove from your home the wireless devices that fail the Love Test.
- ❏ Practice purchasing only items that survive the Love Test.
- ❏ Buy more ethernet cords so you can more easily Disable and Cord.

WiFi router products.

- ❏ Some WiFi routers have been designed to be low-EMF while also performing well with various internet service providers. Since technologies change often, research this to make sure it is the best option for your needs.
- ❏ Router guard. There are small Faraday cages you can put your WiFi router(s) in to reduce EMFs, but be aware that it can slow your connection too. You can find some on LessEMF.com.

OBSERVE YOUR EMF ELIMINATION DIET

- ❏ As you detox your EMFs from your WiFi router(s), record (in Chapter 8 Notes) cause-effect(s) you observe. Many people notice improved sleep quality, and enhanced presence and attention.

EVOLVE YOUR MINDSET

- ❏ Revisit Cornerstone 4 in Chapter 1 to shape your perspective for a lifetime of optimized energy hygiene.

CELEBRATE YOUR PROGRESS!

- ❏ Record your progress in Chapter 1 (Cornerstone 5) to appreciate your growing awareness and achievements. Congratulations, you have learned important ways to detox your sleep!

3.4 HOME ENTERTAINMENT

Many modern home entertainment systems (including speakers, TVs, and projectors), use WiFi and Bluetooth technology. These default settings create chronic EMFs, even when you are not using the devices.

LOVE TEST

- ❑ **Audit.** Take inventory of your home entertainment technology to assess whether they may create Wireless Emissions and assess opportunities to ground their Electrical Emissions.
    - ❑ Desktops.
    - ❑ Laptops.
    - ❑ Game consoles.
    - ❑ Headphones.
    - ❑ Remote controls.
    - ❑ Smart devices.
    - ❑ Speakers.
    - ❑ Televisions.
    - ❑ WiFi boosters.
    - ❑ Besides using a wired connection, go into the settings of the devices checked above to disable WiFi and Bluetooth.
- ❑ **Conduct the Love Test.** Given the health risks from their Wireless Emissions and Electrical Emissions, are there entertainment products that you do not love or need?
- ❑ **Edit your EMF exposure.** Avoid (eliminate, phase out, or don't buy) the home entertainment products that fail the Love Test.
- ❑ **Manage your worthwhile Risks.** Review the Energy Recovery Opportunities below.

CREATE ENERGY RECOVERY OPPORTUNITIES

Decrease your EMF exposure from what survives the Love Test with the strategies below.

- ❑ **Minimize Time.**
    - ❑ Disable WiFi/Bluetooth functions when they are not needed.
    - ❑ If you plan to watch a movie, or listen to music or a podcast on your smart device, then try to download the content before you want to enjoy it/them. And download them when you can be away from the device so you can listen/watch the content when Wireless Emissions are disabled.

- ❑ Assess other entertainment products in your home that use Wireless Emissions, and enable them only when the Wireless Emissions are serving a purpose.
- ❑ **Maximize Distance** from the EMF source.
- ❑ **Create EMF Naps** for home entertainment products that create Wireless Emissions and Electrical Emissions.
  - ❑ If you must use WiFi (like for speakers), then disable their Wireless Emissions and Electrical Emissions either manually or automatically when you do not need them (like at night), and enable them at your preferred wake time.
- ❑ **Ground the Electricity.** Ground as many electrical plugs as is practical to reduce, or stop, the flow of electricity demanded by a product plugged into an electrical outlet. Reference Cornerstone 3 for tips.
- ❑ **Disable and Cord.**
  - ❑ Identify opportunities to disable WiFi and/or Bluetooth connectivity and enjoy your product(s) with an Ethernet cable for wired connections.
- ❑ **Detox the Technology Settings.**
  - ❑ Identify which parts of your home entertainment system can be enjoyed with Wireless Emissions disabled.
  - ❑ Track your research in the Chapter 8 Notes section.

3.5 APPLIANCES

Appliances create EMFs from their electrical currents. But since appliances are quickly being created with WiFi and/or Bluetooth technology, these appliances may also create Wireless Emissions that can often be disabled, and their Electrical Emissions can be reduced.

LOVE TEST

- ❑ Notice the appliances that are plugged in, and that may create Wireless Emissions. When shopping for new appliances, investigate whether they emit Wireless Emissions. Can those Wireless Emissions be disabled/enabled relatively easily to reduce unnecessary emissions? Examples of appliances that can have Wireless Emission are:
  - ❑ Coffee maker.
  - ❑ Dryer.
  - ❑ Freezer.

- Ovens, including a toaster oven. (For example, there are smart ovens that communicate with your cell phone).
- Refrigerator.
- Smart diapers that inform parents when the diaper is wet.
- Thermometers. (For example, thermometers that collect data for your smartphone, or wireless thermometers for cooking).
- Toilets. (For example, there are smart toilets that operate via WiFi or Bluetooth technologies to stream music and other amenities).
- Washing machine.

- **Conduct the Love Test.** Given the health risks from their Wireless Emissions and Electrical Emissions, which appliances do you love or need?
- **Edit your EMF exposure.** Avoid (eliminate, phase out, or don't buy) the appliances that fail the Love Test.
- **Manage your worthwhile Risks.** Review the Energy Recovery Opportunities below.

CREATE ENERGY RECOVERY OPPORTUNITIES

To decrease your EMF exposure from appliances that survive the Love Test:

- **Minimize Time.** If appliances offer capabilities you love that require Wireless Emissions, then assess whether you can disable Wireless Emissions when you are not enjoying the appliances' capabilities.
- **Maximize Distance** from the EMF source and where you spend a lot of time (like where you work).
- **Create EMF Naps** for the EMF source.
  - Which appliances can you create EMF Naps for? Manually or automatically disable Wireless Emissions and/or Electrical Emissions for the nap time, and enable them for the wake time.
  - Check with relevant appliance manufacturers to make sure you understand the potential performance consequences of EMF Naps.
- **Ground the Electricity.** Which appliances can you ground (like with a 3-prong power strip, or grounding adapters) that can reduce, or shut, the flow of electricity demanded by a product plugged into an electrical outlet)?
  - To disable or reduce Electrical Emissions, connect appliances (like coffee makers and countertop ovens) to a 3-pronged power strip, grounding adapter, or timer.
- **Disable and Cord.** Are there appliances whose Wireless Emissions you can disable and still enjoy, or enjoy with an Ethernet cable?

- **Detox the Technology Settings.** Can you disable unnecessary Wireless Emissions in the product's settings?
- Track your findings in the Chapter 8 Notes section.

MICROWAVE AUDIT

While most major authorities deem microwaves safe to use, others voice concerns. To take precautionary measures:

- **Audit.** How often do you use the microwave? How close are people to it when it's in use? How often is it plugged into an electrical outlet? Does it emit Wireless Emissions?
- **Love Test.** Does your microwave pass the Love Test?
- **Edit.** Avoid (eliminate, phase out, or don't buy) microwaves that fail the Love Test. Discard responsibly damaged microwaves.
- **Manage your worthwhile Risks.** If your microwave passes the Love Test, create Energy Recovery Opportunities.
    - Follow the manufacturer's instructions carefully.
    - When your microwave is in use, **Maximize Distance** between you and it.
    - Reduce uses of it.
    - **Create EMF Naps** for its Electrical Emissions. Does it have Wireless Emissions you can disable? If so, **Detox the Technology Settings**.
    - **Ground the Electricity.**

3.6 HOME OFFICE

Our home offices can have many sources of EMF exposure because of the high prevalence of technology products—like lights, computers, printers, routers, and other smart devices.

LOVE TEST

- **Audit.** Notice the EMF sources where you work, and their proximity to where people spend a lot of time. Consider checking the following to identify sources of Wireless Emissions (Cellular/WiFi/Bluetooth) and Electrical Emissions in your home office:
    - Cell phones.
    - Charging stations.
    - Computer mouses.

- ☐ Cordless phones.
- ☐ Desktop computers.
- ☐ Digital screens.
- ☐ Entertainment systems.
- ☐ Fax machines.
- ☐ Lamps.
- ☐ Laptops.
- ☐ iPads.
- ☐ Keyboards.
- ☐ Phones.
- ☐ Power strips.
- ☐ Printers.
- ☐ Scanner machines.
- ☐ Security cameras.
- ☐ Smart devices.
- ☐ Speakers.
- ☐ Televisions.
- ☐ WiFi routers.
- ☐ WiFi boosters.
- ☐ Other wireless devices.

☐ **Conduct the Love Test.** Given the health risks from their Wireless Emissions and Electrical Emissions, are there products that you do not love or need? Which ones do you not mind disabling? Is there a better location for their placement?

☐ **Edit your EMF exposure.** After completing the above, revise your work area to eliminate EMF sources that fail the Love Test.

☐ **Manage your worthwhile Risks.** Review the Energy Recovery Opportunities below.

CREATE ENERGY RECOVERY OPPORTUNITIES

To reduce EMF exposure for the products that survive the Love Test:

- ❏ **Minimize Time.** There will be EMF sources that you cannot avoid. For these, minimize time on, or near, them. Reduce your time to only when the Risks are worthwhile.
- ❏ **Maximize Distance** between EMF sources and where you spend a lot of time.
- ❏ **Create EMF Naps.**
  - ❏ Identify which EMF sources can stay disabled most of the time.
  - ❏ Identify which EMF sources can take EMF Naps.
  - ❏ Create EMF Naps manually or automate them. Cornerstone 3 in Chapter 1 provides more suggestions.
- ❏ **Disable and Cord.** Identify which products contain Wireless Emissions that can be disabled and enjoyed via an Ethernet cable.
- ❏ **Ground the Electricity.** Evaluate plugged-in products and create a plan to ground them. *Be mindful that a computer should be powered off before its power source is disabled to prevent potential damage or disruption.*
- ❏ **Detox the Technology Settings.** Evaluate which products may be able to operate with Wireless Emissions disabled. Examples include most items listed in the preceding Home Office Audit.

OBSERVE YOUR EMF ELIMINATION DIET

- ❏ As you detox your work area's EMFs, record (in Chapter 8 Notes) cause-effect(s) you observe. Many people notice improved sleep quality, energy, presence, and attention.

EVOLVE YOUR MINDSET

- ❏ Revisit Cornerstone 4 in Chapter 1 to shape your perspective for a lifetime of optimized energy hygiene.

CELEBRATE YOUR PROGRESS!

- ❏ Record your progress in Chapter 1 (Cornerstone 5) to appreciate your growing awareness and achievements. Congratulations for detoxing your home work space!

3.7 LIGHT BULBS

Light bulbs can emit unhealthy energy. Below are general guidelines for choosing healthier light bulbs. They are listed from best to worst from an EMF perspective. When selecting which ones to buy for your home, exercise critical thinking, read product labels carefully, and do your own research because "greenwashing" occurs.

- ❏ **Best.** Use incandescent light bulbs if you want to reduce "dirty electricity," a type of energy that EMF-sensitive people prefer to avoid.
  - ❏ From an EMF perspective, incandescent light bulbs are often reported to be the safest ones: they produce less dirty electricity and emit less blue light, which can undermine your sleep quality.

- ❑ However, the tradeoff is that incandescent light bulbs are less efficient than LED and CFL light bulbs.
- ❑ Halogen light bulbs are reported to be a healthy option too. They are less efficient than CFLs, but can be more efficient than incandescent bulbs.

❑ **Second best.** Use light-emitting diode (LED) light bulbs if incandescent light bulbs are not an option. While they tend to be the second-best option (after incandescent light bulbs), they are not necessarily better than CFLs (the worst option).

- ❑ LEDs are popular (including in the screens on smartphones, computer monitors, and televisions) because they last far longer than incandescent bulbs, can be much brighter, run cooler, and use far less electricity.
- ❑ Some LEDs are not unhealthy, and some are unhealthy. Only a meter or third-party testing can know for sure which LEDs are safe. Generally, LEDs emit more blue light, dirty electricity, and EMFs than incandescent light bulbs.
- ❑ If you must use LED bulbs instead of incandescent ones, try to use the ones that do not have a transformer.
- ❑ Beware of dimmable LEDs, which can create dirty electricity.
- ❑ Consider that some white light-emitting diodes (LEDs) emit a wavelength of light associated with adverse human health effects. Consider this when purchasing light bulbs.

❑ **Worst.** Be aware that compact fluorescent light bulbs (CFLs) ("curly pig tail" types of light bulbs) generally create dirty electricity. CFLs can also contain mercury, a neurotoxin that can harm the brain and nervous system. So CFLs create the risk of mercury contamination in your home if the light bulbs break. Upon disposal of CFLs, this mercury remains an environmental health hazard.

- ❑ CFLs are considered the least healthy light bulb type on the market.
- ❑ However, they are popular because CFLs use less energy than incandescent bulbs, and are less expensive than most LEDs.
- ❑ If CFLs pass your Love Test, or you have a pre-existing inventory of them you want to use up, or you must use them at work, try to use ones that are "double encapsulated." If the CFLs accidentally break, this double encapsulation helps to avoid, or minimize, mercury from leaking out.

❑ Consider using nightlights with dim red light bulbs, as red light has the lowest effect on melatonin production and circadian rhythms.

❑ Minimize exposures to unnecessary artificial light. Scientists have raised health concerns from our exposures to other types of artificial light too. For example, those exposed to lots of cool white fluorescent lights may have higher risks of developing cataracts.

LIGHT BULB HELP

- ❑ Print an image of the different light bulbs mentioned above and staple it to this page, or the Chapter 8 Notes section, to help your shopping selections. You can find an image of incandescent, LED, CFLs, and halogen light bulbs in the online EMF Detox workshop at the D-Tox Academy in lesson 3.7. Alternatively, you can search online.

- ❑ Read my article "Could dirty electricity be affecting your health?" that I wrote for Well+Good. You can search for it online. Members can find this linked in lesson 3.7 in the online EMF Detox workshop.

3.8 EMF EXPERT

While everything in EMF Detox will help you hack your EMFs, you may be interested in reducing your EMF exposure further. You can buy a meter to identify EMF sources in your home that are easy for you to detox (like overlooked wireless devices), but meters can be hard to make sense of.

Regardless, hiring a trained EMF expert to measure your home will undoubtedly reveal additional ways to reduce your EMF exposure, if that is an option for your budget and geography.

Everything we have reviewed so far will prepare you to get even more value from a hired EMF assessment professional. If hiring a professional is not an option, these checklists will help tremendously so just keep incorporating them into your habits.

To help find a trained EMF assessment professional:

- ❏ Look for an EMF expert certified by the Building Biology Institute: https://buildingbiologyinstitute.org. The Building Biology Institute EMRS Program is where the best EMF experts learn how to test and remediate electromagnetic radiation and get certified. The EMRS Certification is considered the most rigorous and comprehensive electromagnetic training and education program in the world.

CERTIFIED EMF EXPERT

I recommend James Finn, founder of Elexana LLC. Based in New York City, Elexana has evolved from being the first official EMF Testing Company to an industry-leading U.S. National Electromagnetic Services Company. Elexana LLC tests, analyzes, and mitigates EMF radiation to attenuate electromagnetic interference (EMI.) Learn more at https://www.elexana.com.

3.9 TECH HOME

Establishing a **"Tech Home"** (aka, a charging station for your home's technology) is a great way to get a household to organize all technology into one place by a certain hour. Other great benefits can be enjoyed by following the Tech House Rules below.

"TECH HOUSE RULES"

Using a Tech Home mitigates many challenges (including social-emotional ones) if these requests are followed:

- ❏ Establish the Tech Home at least 6-8 feet from where people spend a lot of time, like where they work.
- ❏ Definitely locate the Tech Home at least 8 feet from where people sleep. Remember that EMFs can travel through walls, floors, and ceilings.
- ❏ Organize all chargers into a power strip.
    - ❏ Disable Electrical Emissions of the power strip when not used.
- ❏ Develop a "bedtime" by which all laptops and smart devices (and their related chargers and accessories) must return to the Tech Home.
- ❏ **Create EMF Naps** for your technology (follow tips to disable Wireless Emissions in Chapter 1), before plugging them into their chargers in the Tech Home.
- ❏ **Ground the Electricity.** Follow tips to disable Electrical Emissions in Chapter 1.

—4—

EVENING EMF EDIT

Be mindful of your environmental stimulants.

The online videos in EMF Detox highlight common evening exposures that can spoil your rest. As you learn more about how stimulants—like blue light—can undermine your sleep quality and how other overlooked exposures can distract from healthy sleep habits, discover new evening rituals. To spend your evening in ways that intentionally soothe you for sleep can be life-changing. Enjoy the discovery process!

Edit your evening routines to incorporate this section's tips. Doing so can re-balance your time, body, soul, and mind for deeper restoration.

4.1 BLUE LIGHT

Thanks to our circadian rhythms, our bodies are primed to be awake with light, and asleep with darkness. Blue light—from the screens of phones, iPads, computers, TVs, light bulbs, and more—can interfere with our circadian rhythms, trigger our body's wakefulness responses, and make it harder to sleep well. To be mindful of your blue light exposures:

- ❑ Check EMF Detox online for the latest recommendations in which apps or software to download and use to filter light from digital screens, including blue light or "flickering."

- ❑ Remember that exposure to natural bright light during the day improves your ability to sleep at night, but that light is generally stimulating so you need to be mindful of your light exposures later in the day and in the evening.

- ❑ Enable "Night Shift" on your wireless devices and computer screens. For example, many phones and laptops offer a functionality that shifts display colors from bright blue to a warmer end of the spectrum at sunset.

- ❑ Avoid looking at bright screens 2-3 hours before bedtime.

- ❑ Wear blue light glasses that filter the blue/green wavelength at night. Recommendations are in section 4.1 in the online EMF Detox workshop.

- ❑ After "disconnecting" from devices for the night, choose a soothing activity before bedtime (preferably not on a digital screen). Even 5-15 minutes of resetting your mind, mood, and eyes before putting your head on the pillow may ease your transition into restful, restorative sleep.

4.2 YOUR BEDROOM

Our bedroom should be a place of respite, where our bodies can repair themselves while we sleep. However, due to an increasing amount of technology in our bedrooms, our technology can also create chronic sources of EMF exposure, which can undermine restorative sleep.

Considering we spend 1/3 of our lives sleeping, detoxing your bedrooms' EMFs is high impact. So prioritize your EMF Detox efforts for where people sleep.

NIGHTSTAND EMF AUDIT

- ❑ **Audit.** Assess what is around your bed for Wireless Emissions and Electrical Emissions (what are plugged into nearby electrical outlets). Common EMF sources on night stands, or near beds, include:
    - ❑ Alarm clocks. A better EMF choice is using a battery-powered alarm clock instead of a corded one.
    - ❑ Cordless phones.
    - ❑ Cell phones.
    - ❑ Diffusers.
    - ❑ Phone chargers.
    - ❑ Lamps.
    - ❑ Smart devices.
    - ❑ Stereo.
- ❑ **Conduct the Love Test.** Given the energetic stressors that may be on your nightstand or near your bed, which do you not need or love?
- ❑ **Edit your EMF exposure.** After completing the above, revise what remains near your bed, and on your nightstand.
- ❑ **Manage your worthwhile Risks.** Review the Energy Recovery Opportunities below.

CREATE ENERGY RECOVERY OPPORTUNITIES

To reduce your EMF exposure from what passes the Love Test:

- ❑ **Maximize Distance** between where you sleep and EMF sources. Alarm clocks, lamps, and other EMF sources that pass the Love Test can be located as far as possible from where you sleep, like on a dresser across the room.
- ❑ **Create EMF Naps** for your Wireless Emissions.
    - ❑ Set up manual or automated ways to disable and enable Wireless Emissions and (maybe) Electrical

Emissions at specified times.

- ☐ If you will do this manually, make this a part of your evening and morning routines.
- ☐ Or, automate this with a timer to have the WiFi router turn on/off automatically at set times.

☐ **Ground the Electricity.** If products that pass the Love Test create Electrical Emissions, Ground the Electricity with a grounding adapter (an example is on Nontoxic Living on Amazon at https://www.NontoxicLiving.tips), which allows you to reduce, or stop, the flow of electricity, including with a remote control. Refer to Chapter 1 for tips.

☐ If relevant, prioritize the following checklists for where you sleep:

- ☐ 2.1 Airplane Mode.
- ☐ 2.6 Laptops.
- ☐ 3.1 Power Strips.
- ☐ 3.2 Cordless Phones.
- ☐ Any other EMF sources.

EVENING RITUALS EMF EDIT

As a reminder from prior checklists:

☐ Establish a bedtime for your technology. At least one hour (or more) before bedtime, power off computers, tablets and phones. Keep all devices you do not need out of your bedroom.

☐ Create and use a technology "home," or charging station. This can organize all technologies and related items. As part of this evening routine, disable Wireless Emissions and charge devices for the next day. Reference checklist 3.9 for Tech House Rules.

☐ Develop replacement evening activities that make it easier—even enjoyable—to have non-tech or low-tech evenings.

SLEEP EMF AUDIT

☐ Avoid sleeping with wireless devices near you. If they must be near you, disable Wireless Emissions before you sleep.

☐ Avoid or minimize using electric blankets, water beds, and electric heating pads.

☐ Leave at least eight inches of space between your bed and the wall because wiring (even in walls) can still emit a significant magnetic field you do not want to be sleeping in.

☐ If your bedroom is located right next to a utility pole, position your bed so it is at least six feet away from this strong magnetic field.

EVENING EMF EDIT

- ❑ Since the refrigerator and home entertainment center are often the biggest producers of magnetic fields in the home, consider moving your bed (or bedroom) to be as far from them as possible.
- ❑ If your bedroom is next to the kitchen or the home theater, consider moving your bed to the opposite side of the room to establish a safe distance.
- ❑ Metal may contribute to EMFs. Sensitive people may sleep better with less metal around the bed.
- ❑ Before purchasing a new mattress, assess the metals in the mattresses. There is a theory that metal innersprings may contribute to a heightened EMF. This may not affect most people. But, if you are hypersensitive, you may prefer mattresses without metal.
- ❑ Avoid electric beds, as these may create higher EMFs.
- ❑ Try to sleep in the dark. If you cannot create a dark room (like with blackout shades), then consider sleeping with a nontoxic eye mask to help protect your sleep from light. You can see which one I use on **Nontoxic Living on Amazon**.
- ❑ Review the Home Detox checklists at the D-Tox Academy or in the *Home Detox Workbook* to prioritize those recommendations for your bedroom. They will lead you to detox chemicals and heavy metals in your home and habits.
- ❑ Explore with a certified EMF professional whether switching off the circuit breaker at night can lead to more restorative sleep. If this could help you, then you may want to alter your bedroom circuit breaker to exclude necessities—like a smoke detector, alarm, refrigerator.

OBSERVE YOUR EMF ELIMINATION DIET

- ❑ As you detox your bedroom's EMFs, record (in Chapter 8 Notes) cause-effect(s) you observe. Many people notice improved sleep quality, and enhanced presence and attention.

EVOLVE YOUR MINDSET

- ❑ Revisit Cornerstone 4 in Chapter 1 to shape your perspective for a lifetime of optimized energy hygiene.

CELEBRATE YOUR PROGRESS!

- ❑ Record your progress in Chapter 1 (Cornerstone 5) to appreciate your growing awareness and achievements. Congratulations for detoxing your bedroom, where you spend approximately 1/3 of your life!

— 5 —

BODY EMF EDIT

Be curious about our unique bodies and the universes within.

The Body EMF Edit checklists plant seeds of curiosity for so much more to discover about how unique each of us is: parent from child, sister from sister, brother from sister, parent from grandparent, etc. And while we each have unique vulnerabilities (though generalizations—like children and pregnant women—are helpful too), we also have unique healing stimulants.

As you complete the EMF Detox checklists, I hope you remember that you have just begun your journey for practical nontoxic living and healing. And I hope you arrived at this stage with amazement and curiosity to explore it further.

5.1 UNIQUE VULNERABILITIES

It is important to remember that our biology is unique. But generalizations sometimes help: older adults tend to have different vulnerabilities than younger adults; pregnant women are more susceptible to certain exposures than non-pregnant women; men and women have different vulnerabilities from each other; and so forth.

In addition to applying the other checklists in this workbook with extra diligence for those with unique vulnerabilities, evaluate the below items for potential Wireless Emissions or Electrical Emissions:

- ❑ Baby Monitors. Some can emit risky EMFs. Evaluate the risks/rewards of using them. Remember that the further away the baby monitors are from the baby, the lower the EMFs. Better products enter the market each year so research for a low EMF baby monitor if you need one.
- ❑ Smart diapers.
- ❑ Smart watches.
- ❑ Smart thermometers.
- ❑ Security cameras.
- ❑ Toys.

Additional tips or reminders:

- ❑ Children are particularly vulnerable. Unborn children in women are even more vulnerable. However, stress management is as important a pillar of health as any so prioritize your stress management as you apply the EMF D-Tox Approach.
- ❑ **Minimize Time.** While it is ideal to have children spend no time around Wireless Emissions or Electrical Emissions, that is often impossible. Instead, do your best to minimize how much time they spend near them. Tips:
  - ❑ Do not allow cell phones to be used as toys or teething items.
  - ❑ Do not allow wireless devices in children's bedrooms.
  - ❑ Enforce the Tech Home as a family ritual (discussed in checklist 3.9 Tech Home).

"Children are disproportionately affected by environmental exposures, including cell phone radiation. The differences in bone density and the amount of fluid in a child's brain compared to an adult's brain could allow children to absorb greater quantities of RF energy deeper into their brains than adults. It is essential that any new standards for cell phones or other wireless devices be based on protecting the youngest and most vulnerable populations to ensure they are safeguarded through their lifetimes."

—Thomas K. McInerny to the Honorable Dennis Kucinich, December 12, 2012,

on behalf of the American Academy of Pediatrics

- ❏ If children are using an internet-enabled phone, or other device for work or play, try to plan their use of it with Wireless Emissions disabled.
- ❏ **Maximize Distance.**
  - ❏ Minimize use of wireless devices while a child is on your lap, in your arms, or in your belly. When using wireless devices, keep as much distance from them as possible. For example, use technology while technology rests on a table rather than on your body. Whenever possible, use them with Wireless Emissions and Electrical Emissions disabled (check Chapter 1 for tips).
- ❏ Consider what other materials (like metal) may contribute to creating an EMF field around the child.
  - ❏ Assess how much metal is where the child sleeps. The risks from metal is unknown.
  - ❏ Some mattresses contain metal springs. Most scientists are not concerned about metal springs in mattresses but some people prefer to avoid them.
- ❏ When children use digital devices, try to convince (or inspire) them to wear glasses that filter blue light.
  - ❏ Consider checklist 4.1 Blue light for how to protect your uniquely vulnerable.
- ❏ Besides young life, other demographics that are more vulnerable include the elderly, others with compromised health, and those with multiple chemical sensitivity. For them, consider pursuing precautionary measures more diligently.

5.2 MEDICAL IMAGING

Though medical imaging (like X-Rays) can be life-enhancing and even life-saving, the growth in medical imaging has skyrocketed. The tips that follow are meant to encourage you to collaborate with your trusted physicians to minimize medical imaging responsibly, and to take precautionary measures to decrease unnecessary radiation exposures.

To reduce unnecessary EMF exposures from medical imaging, when you or your family need them:

- ❏ Talk to your trusted healthcare providers about an estimated schedule of needed medical imaging (the number and frequency of images), and create a plan you both feel is medically responsible and will avoid unnecessary EMF exposures; i.e., you and your physicians should discuss the benefits and the risks of the medical imaging plan.
- ❏ If you are trying to make medical imaging decisions with a specialist or in the emergency room, consider reaching out to your trusted physician (like your general practitioner or pediatrician) to help you navigate medical imaging decisions. Be aware of how much this may increase your medical expenses but consider this risk/reward option.
- ❏ If you are making medical imaging decisions for your children:

- ❏ Visit this website to learn precautionary measures you may be able to pursue or discuss with trusted healthcare providers: http://www.imagegently.org.

- ❏ When the pediatric dentist suggests x-rays, discuss the risks/rewards of medical imaging (like x-rays) in consideration of the child's age, the benefits of the medical imaging, and how to responsibly minimize the number of x-rays over the course of years and spread them out over time. Pediatric dentists have begun x-rays at an earlier age, so children experience more x-rays than ever. By explaining concerns to your dentist, your dentist may be able to find alternative solutions.

- ❏ When medical imaging is being performed, be sure the patient's body is as protected as it can be. Ask whether the patient could be more covered, or more protected, with radiation-protective shields.

- ❏ When medical imaging is pursued, talk to your physicians and medical imaging technicians about whether the machines' settings are set to expose the patient to minimal radiation without compromising the usefulness of the images.

- ❏ Assess opportunities to have physicians collaborate on medical images. *I have had to work with a couple of different dentists who specialize in different things. Each wanted their own set of dental x-rays. I had to persist in having the dentists collaborate on the medical imaging they each wanted. When they could share the same set, my radiation exposure was reduced.*

5.3 EARTHING

It is important to remember that EMFs are not necessarily bad. Both the earth and our bodies produce their own electromagnetic fields. Some studies show benefits from human exposure to natural electromagnetic fields of the earth and from others.

Our modern lifestyles, however, have created more disconnection from nature because we spend more time indoors. Research suggests that reconnecting with nature—especially skin contact with the earth—enhances our well-being.

To experience "earthing" or "grounding" with the earth's frequency:

- ❏ Walk barefoot on the grass. *Grass with some water (like dew) can be even better.
- ❏ Hug a tree. Literally!
- ❏ Spend time amongst trees (aka, "forest bathing"). A great way to do this is by hiking among trees.
- ❏ Walk on the beach with your bare feet on the sand.
- ❏ Carve out time in your schedule to connect with the Earth regularly.
- ❏ Garden.
- ❏ Adopt indoor plants.

❑ If you do not have time to do any of the above, touch plants, especially the stems, as they provide opportunities to "earth."

EARTHING & GROUNDING PRODUCTS

It is hard to know which EMF protection products truly help. Many can cause more harm than good. However, the guest of *Practical Nontoxic Living* podcast #17 is ex-telecom engineer and executive Daniel DeBaun, author of *Radiation Nation*. Listen to it to hear his thoughts on EMF protection products that he developed to protect his sons.

5.4 NATURAL REMEDIES

Generally, we should seek practical ways to protect, unburden, and support the brain, nervous system, cardiovascular system, endocrine system, immune system, and our genetics. These efforts should help the body's reactions to EMF exposure.

The list below offers suggestions that may not necessarily be supported by science (generally, no one is funding the study of natural remedies to counter the stressful effects from EMF exposures), but the suggestions should pose little to no health risks, while offering benefits.

Natural remedies aim to reduce stress, anxiety, exposures to toxic chemicals/heavy metals/EMFs, and negativity. They can also help boost detoxification, healing, immune responses, sleep quality, energy, presence, feel-good hormones, social and emotional connections, and positivity.

To support your body's resiliency, consider:

- ❏ Baths. While a hot bath can be draining for some, baths can be relaxing and even detoxifying too. Nontoxic, detoxifying options include baths with sea salt and baking powder (various recipes are online but one includes one pound of each), organic flower petals, organic herbs, essential oils (if you don't experience adverse reactions), coconut milk, or even just water.
    - ❏ It may help to submerge as much of the body as possible for twenty to fifty minutes.
    - ❏ Hydrate frequently with filtered water.
    - ❏ When you are done, drain the tub and rinse the baking soda, salt, and pollutants from you and then your tub.
    - ❏ Use the Chapter 8 Notes section to track bath recipes that may counter the effects of radiation exposure.
- ❏ Detox your home, diet, and self-care from toxic chemicals and heavy metals. My online Home Detox workshop and *Home Detox Workbook: Checklists to Eliminate Toxic Chemicals* are very helpful!
- ❏ Exercise regularly. Accessible routines can include a relaxing walk outdoors, if it is safe (consider the current infectious disease recommendations), or exercise to an online video. Many are free on YouTube and Instagram. Affordable ones are available on many apps too.
- ❏ Spend time in nature. Check tips in checklist 5.3 Earthing.
- ❏ Crystals that may protect against stressful EMFs include Shungite, Black Tourmaline, and Lepidolite. However, there have not been peer-reviewed scientific analyses on these claims.
- ❏ Get a healthy dose of natural sunlight for vitamin D production as supplements may or may not help.
- ❏ Have a great conversation with someone you care about.
- ❏ Help a community that is meaningful to you.

- ❏ Take digital detoxes, if even for just an evening or a morning. The longer they can be, the more benefits you can experience.

- ❏ Optimize your diet. Eat a diet that empowers your body's natural detox and healing processes, and that provides minimal burdens (i.e., minimize eating processed foods, excess sugar, and alcohol). While what is an optimal diet for one person is not necessarily optimal for another, generally, an organic, plant-based, whole foods diet is very beneficial. The suggestions below have been reported to help the body counter the adverse effects from radiation exposures, but the science may not always back it up (few studies have been pursued on them). The list is limited to foods that may both mitigate damaging effects of radiation and offer other health benefits. The intention is that they should **cause no harm** and may provide benefits. This list is meant to jumpstart your own research to explore if they make sense for you. If you have health sensitivities, please discuss with your trusted healthcare professionals the incorporation of new foods.

    - ❏ Bee pollen.
    - ❏ Brewer's yeast.
    - ❏ Beets.
    - ❏ Anti-inflammatory foods or ingredients, like curcumin.
    - ❏ Genistein, which is often found in soy products.
    - ❏ Miso (fermented soybean paste). Doctors in Japan notice that miso seems to help heal those who suffered from radiation poisoning.
    - ❏ Other nutrient-dense foods to boost your immunity, blood quality, nervous system, and detoxification.

- ❏ Negative ions' exposure may help. Our technology (like appliances and computers) can emit positive ions, so our indoor environments can become imbalanced and outnumbered by positive ions. To balance your ions:

    - ❏ Minimze Electrical Emissions (this may decrease indoor positive ions, which may burden the body).
    - ❏ If outdoor air quality is good, open the windows (it may help balance the ions indoors).
    - ❏ Take a shower (it can expose you to healthy negative ions).
    - ❏ Improve indoor health through plants (plants may consume positive ions, which is good).
    - ❏ Spend time outdoors in areas with high negative ions (which may help counter the negative effects from lots of technology), such as the seaside, mountains, forests, and having skin contact with the earth.
    - ❏ Decrease things in your home that are made of synthetic materials, like plastics, polyester, etc.
    - ❏ Conduct the Love Test frequently, and edit your things accordingly. Remember that having just what you need and love is more healthy for body, mind, and energy.

LISTEN TO SCIENCE-BASED APPROACHES TO HEAL

- ❑ Incorporate nature into your home through plants, images of nature, and natural materials (like natural wood, stones, or crystals). The sights and smells of nature can promote healing. Learn more from the *Practical Nontoxic Living* podcast #18 with Dr. Esther Sternberg, MD, author of *Healing Spaces: The Science of Place and Well-being* and *The Balance Within: The Science Connecting Health and Emotions*.

- ❑ Meditate or discover a mindfulness practice. Listen to *Practical Nontoxic Living* podcast #23, Meditation for Beginners, with guest Tal Rabinowitz, founder of Den Meditation for help.

- ❑ Consider your human electromagnetic field. Author of *Energy Medicine: The Science and Mystery of Healing*, Dr. Jill Blakeway shares fascinating insight on this in *Practical Nontoxic Living* podcast #25.

- ❑ Detox your home of toxic chemicals and heavy metals. The online Home Detox workshop at https://NontoxicLiving.tips and *Home Detox Workbook: Checklists to Eliminate Toxic Chemicals* are very helpful.

5.5 DIGITAL DETOXES

The prior checklists help you identify detox tweaks. A more intentional EMF "vacation" can be even more enlightening and healing. When you feel motivated to try this, experiment with longer digital detoxes to observe how you are impacted. **Key to a successful digital detox is developing alternatives habits that are enjoyable.**

BLUETOOTH DETOX

Some people are sensitive to Bluetooth but do not realize it yet. Experiment with taking breaks from Bluetooth to see if you notice effects. If you are sensitive, you may notice the difference when re-exposed to Bluetooth after a Bluetooth detox. For example, some people notice improvements in sleep, headaches, fatigue, nausea, or other symptoms.

- ❑ After identifying your Bluetooth sources, select a day, night, weekend, week, or month to decrease your Bluetooth exposures.
- ❑ Use the Chapter 8 Notes section to track your symptoms and sensations before and after a Bluetooth detox.

CELL PHONE DETOX

Explore how you may be changed by spending less time connected to your cell phone.

- ❑ Select an evening, weekend, or more to live without your cell phone.
- ❑ Use the Chapter 8 Notes section to track how you feel before and after a cell phone detox.

WIFI ROUTER DETOX

Some people sleep and feel better from disabling their WiFi router(s), especially at night. To learn your body's response to your WiFi router emissions, experiment with disabling your WiFi router(s)' Wireless Emissions and Electrical Emissions during the times below.

- ❑ When sleeping.
- ❑ When at home during the day for long periods of time.
- ❑ As a default setting.
- ❑ Use the Chapter 8 Notes section to track how you feel before and after eliminating an EMF source.

—6—

CONGRATULATIONS!

All your baby steps have led you to so much EMF awareness and more recovery opportunities. You are well positioned to continue enjoying even more EMF recovery times—and from using your intuitive common sense to make EMF detox tweaks. Please reference these checklists when you are ready for reminders or more tips.

HOORAY! YOU DID IT!

CONGRATULATIONS!

FEEL PROUD OF YOURSELF!

—7—

MY SOCIAL INSPIRATIONS

Help spread awareness and inspiration.

Share inspiring revelations on social media to help raise awareness and prompt productive public dialogue. Your experience can teach and motivate others! To help, use the template on the next page.

If you are comfortable sharing this on Instagram and Facebook, please tag me: @ruanliving.com. (Instagram is the platform I use more often.) Please use hashtags: #nontoxicliving #ruanliving and #lovetest when relevant.

If you prefer, feel free to email me at hello@nontoxicliving.tips for me to consider sharing to my community.

Random submissions can win a prize!

WHAT I WISH MORE PEOPLE KNEW ABOUT EMFs:

@ruanliving #nontoxicliving #ruanliving

—8—

NOTES

Use this section to track your EMF elimination diet and more.

Examples of what to take notes on are below.

- ❑ Key things to look for in certain products, e.g., which light bulbs to avoid and which to buy, grounding adapters, and WiFi routers that emit less risky EMFs.
- ❑ Results of comparison shopping.
- ❑ Favorite products.

Tips:

- ❑ Write in **pencil or erasable pen** to maximize flexibility in editing your notes. **Color code** your notes!
- ❑ Add the key topics in this Chapter 8 and their starting page numbers to the **Table of Contents** so that you can more easily access your key notes. Use **fun paper clips** with tassels or pom poms to tag key topics you want to access often.
- ❑ Feel free to **staple or clip additional pages** as needed. **Sticky notes** are welcome!

OBSERVATIONS FROM YOUR ELIMINATION DIET

Notice cause-effect relationships from EMF exposures.

1. To start, document your Baseline Assessment by completing the following:
   - ❑ List what you circled in Cornerstone 1 in Chapter 1.
   - ❑ Visit the websites listed in Cornerstone 1 for more information on potential health effects, and add relevant issues in this section.
   - ❑ Add any other chronic health issues that affect you.
   - ❑ Add start dates of an EMF elimination.
2. Record changes in your symptoms, sleep, energy, emotions, mental clarity, etc. as you incorporate changes from the checklists. Note relationships between your reduced EMF exposures and your symptoms.

EMF DETOX

NOTES

EMF DETOX

NOTES

EMF DETOX

NOTES

EMF DETOX

NOTES

EMF DETOX

NOTES

EMF DETOX

NOTES

NOTES

EMF DETOX

NOTES

EMF DETOX

NOTES

EMF DETOX

NOTES

APPENDIX

Select References

This workbook results from many years of research, *Practical Nontoxic Living* podcast interviews, and assessments from EMF professionals who have remediated my homes. Listed below are works cited and resources to help you learn more about EMFs and their complex influences on human health.

5G Appeal 2018. "The 5G Appeal." 2018. http://www.5gappeal.eu.

Belyaev et al. 2016. Belyaev, Igor, Amy Dean, Horst Eger, Gerhard Hubmann, Reinhold Jandrisovits, Michael Kundi, Hanns Moshammer, et al. 2016. "EUROPAEM EMF Guideline 2016 for the Prevention, Diagnosis and Treatment of EMF-Related Health Problems and Illnesses." Reviews on Environmental Health 31 (3): 363–97. https://doi.org/10.1515/reveh-2016-0011.

Bioinitiative Report 2012. "BioInitiative Report 2012." 2020. https://bioinitiative.org/.

California Medical Association House of Delegates Resolution Wireless Standards Reevaluation 2014. Resolution 107- 14. Date Adopted Dec 7, 2014. https://ehtrust.org/the-california-medical-association-wireless-resolution/.

Children's Health Defense 2020. "Robert Kennedy, Jr.'s Legal Team Sues FCC over Wireless Health Guidelines." 2020. https://childrenshealthdefense.org/news/robert-kennedy-jr- assembles-legal-team-to-sue-fcc-over-wireless-health-guidelines/.

Davis, Devra. *Disconnect: The Truth About Cell Phone Radiation, What the Industry Has Done to Hide It, and How to Protect Your Family*. Environmental Health Trust. November 21, 2013.

Dewey, Caitlin. "Are 'WiFi allergies' a real thing? A quick guide to electromagnetic hypersensitivity." Washington Post. 2015. https://www.washingtonpost.com/news/the-intersect/wp/2015/08/31/are-wifi-allergies-a-real-thing-a-quick-guide-to-electromagnetic-hypersensitivity/.

Environmental Health Trust 2016. "Telecom And Insurance Companies Warn Of Liability And Risk." 2016. https://ehtrust.org/key-issues/cell-phoneswireless/telecom-insurance- companies-warn-liability-risk-go-key-issues/.

Environmental Health Trust 2019. "Insurance Authorities Rate 5g And Electromagnetic Radiation As "High Risk"." 2019. https://ehtrust.org/key-issues/reports-white-papers- insurance-industry/.

Environmental Health Trust 2020a. "Worldwide Action On Wi-Fi And Electromagnetic Radiation In School." 2020. https://ehtrust.org/health-effects-wireless-in-schools/.

Environmental Health Trust 2020b. "Alabama Proclamation On Electromagnetic Sensitivity 2020." 2020. https://ehtrust.org/alabama-proclamation-electromagnetic-sensitivity/.

Environmental Health Trust 2020c. "Russia Bans Wi-Fi And Smartphones For Distance Learning." 2020. https://ehtrust.org/russia-bans-wi-fi-and-smartphones-for-distance-learning/.

Environmental Health Trust 2020d. "Medical Doctors And Public Health Organizations." https://ehtrust.org/science/medical-doctors-consensus-statements-recommendations-cell-phoneswireless/ (accessed July 27, 2020).

European Environment Agency 2002. "Late Lessons from Early Warnings: The Precautionary Principle 1896-2000." 2002. https://www.eea.europa.eu/publications/environmental_issue_report_2001_22.

European Environment Agency 2019. "Radiation Risk from Everyday Devices Assessed." 2019. https://www.eea.europa.eu/highlights/radiation-risk-from-everyday-devices-assessed.

Gushée, Sophia Ruan. 2018. "Headaches, Nausea, and Fatigue. Might You Be Electrohypersensitive?" *Practical Nontoxic Living* podcast #15. 2018. https://www.nontoxicliving.tips/blog/brain-cancer-emfs-emf-protection-products-vaccines- and-more-with-dr-david-o-carpenter.

APPENDIX

Gushée, Sophia Ruan. 2019. "5G Rollout: How to protect our health from this new type of EMF radiation" *Practical Nontoxic Living*.podcast #17. https://www.nontoxicliving.tips/blog/5G-rollout-what-can-we-do-to-protect-our-health-from-this-new-emf-radiation.

Gushée, Sophia Ruan. 2020. "Protect Your Brain and Body from 5G and Other EMFs." *Practical Nontoxic Living* podcast #24. 2020. https://www.nontoxicliving.tips/natural-healthy-living-podcast.

Hedendahl 2015. Hedendahl, Lena, Michael Carlberg, and Lennart Hardell. 2015. "Electromagnetic Hypersensitivity – an Increasing Challenge to the Medical Profession." Reviews on Environmental Health 30 (4): 209–215. https://doi.org/10.1515/reveh-2015-0012.

Hippocrates Electrosmog Appeal Belgium 2020. Last updated: June 28, 2020. https://en.hippocrates-electrosmog-appeal.be/signataires.

Hoffman, Chris. 2020. "What Is 5G, and How Fast Will It Be?" How-To Geek. 2020. https://www.howtogeek.com/340002/what-is-5g-and-how-fast-will-it-be/.

Johansson, Olle. 2015. "Electrohypersensitivity: A Functional Impairment Due to an Inaccessible Environment." Reviews on Environmental Health 30 (4): 311–21. https://doi.org/10.1515/reveh-2015-0018.

Kalmbacher, Colin. 2020. "Scientists Sue FCC for Dismissing Studies Linking Cell Phone Radiation to Cancer." Law & Crime. 2020. https://lawandcrime.com/administrative-law/scientists-sue-fcc-for-dismissing-claims-that-cell-phone-radiation-causes-cancer/.

McInerny, Thomas K. 2012. Thomas K. McInerny to The Honorable Dennis Kucinich, December 12, 2012, on behalf of the American Academy of Pediatrics, https://ehtrust.org/wp-content/uploads/American-Academy-of-Pediatrics-Letters-to-FCC-and-Congress-.pdf.

Parliamentary Assembly of the Council of Europe 2011. "The potential dangers of electromagnetic fields and their effect on the environment." Resolution 1815 (2011). http://assembly.coe.int/nw/xml/XRef/Xref-XML2HTML-en.asp?fileid=17994&.

Science for Environment Policy (2017) *The Precautionary Priniple: decision making under uncertainty*. Future Brief 18. Produced for the European Commission DG Environment by the Science Communication Unit, UWE, Bristol. Available at: http://ec.europa.eu/science-environment-policy.

Spangler, Todd. 2019. "U.S. Households Have an Average of 11 Connected Devices — And 5G Should Push That Even Higher." Variety. https://variety.com/2019/digital/news/u-s-households-have-an-average-of-11-connected-devices-and-5g-should-push-that-even-higher-1203431225/.

State Of New York Public Service Commission 2019. "Post-hearing Brief Of Intervenor D. Kopald." 2019.http://documents.dps.ny.gov/public/Common/ViewDoc.aspx?DocRefId=%7B41C3548B-A6B0-4D44-9C78-7B2757F54178%7D.

Physicians for Safe Technology 2016. "Digital Technology and Public Health." 2016. https://mdsafetech.org/..

Reuben, Suzanne H. 2010. "Reducing Environmental Risk: What We Can Do Now." President's Cancer Panel: 2008–2009 Annual Report. 2010. https://deainfo.nci.nih.gov/Advisory/pcp/annualReports/pcp08-09rpt/PCP_Report_08-09_508.pdf.

United States Access Board 2006. "Recommendations for Accommodations." 2006. https://www.access-board.gov/research/completed-research/indoor-environmental-quality/recommendations-for-accommodations.

US Federal Communications Commission 2019. "Wireless Devices and Health Concerns." Date Last Updated/Reviewed: Tuesday, October 15, 2019. https://www.fcc.gov/consumers/guides/wireless-devices-and-health-concerns.

US Food and Drug Administration 2020. "Do Cell Phones Pose a Health Hazard?." Content current as of: 02/10/2020. https://www.fda.gov/radiation-emitting-products/cell-phones/do-cell-phones-pose-health-hazard#:~:text=Scientific%20Consensus%20on%20Cell%20Phone%20Safety&text=The%20weight%20of%20nearly%2030,rates%20in%20the%20U.S.%20population (accessed July 2020).

US Government Accountability Office 2012. "Exposure and Testing Requirements for Mobile Phones Should Be Reassessed." https://www.gao.gov/products/GAO-12-771.

World Health Organization 2005. "Electromagnetic Fields and Public Health: Electromagnetic Hypersensitivity." 2005. https://www.who.int/peh-emf/publications/facts/fs296/en/.

APPENDIX

Detox Deep Dive

EMF Detox Workbook is the second in the *Detox Deep Dive* series. Be sure to consider *Home Detox Workbook: Checklists to Eliminate Toxic Chemicals*. They are essential pillars of a healthy home and lifestyle. The tips were curated to strike the perfect balance of offering maximum impact on your home and lifestyle detox and minimum impact on your budget. Below are others in this series that you can enjoy.

A to Z of D-Toxing: The Ultimate Guide to Reducing Our Toxic Exposures

"Her in-depth research makes this book a helpful, easy-to-ready manual for any household concerned with reducing toxins in everyday life." —Frank Lipman, MD, *New York Times* bestselling author of *The New Health Rules*

"I am truly impressed with Sophia Gushée's book. It is an unbelievable resource... I will turn to it often in my clinical practice." — Hooman Yaghoobzadeh, MD, one of America's best doctors according to *New York* magazine and Castle Connolly, NY Presbyterian Hospital Weill Cornell Medical Center

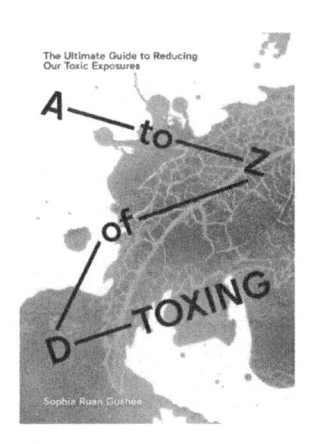

Stay updated!

Be sure to register for Sophia's email newsletter to stay updated on her newest workbook, events, podcasts, and for free practical nontoxic living and healing tips. Just text "DETOX" to the number 66866 to subscribe.

Reach out!

Extreme dedication went into making this workbook be the most responsible and best offering for you. Inevitably, things may get overlooked. Feedback, corrections, and updates are welcome at hello@nontoxicliving.tips!

Made in the USA
Coppell, TX
19 December 2021